Analysis and Design of Steel Structures

Analysis and Design of Steel Structures

Adalberto West

LANRYE
INTERNATIONAL
www.clanryeinternational.com

Clanrye International,
750 Third Avenue, 9ᵗʰ Floor,
New York, NY 10017, USA

ISBN: 978-1-64726-643-1

Cataloging-in-Publication Data

Analysis and design of steel structures / Adalberto West.
    p. cm.
Includes bibliographical references and index.
ISBN 978-1-64726-643-1
1. Building, Iron and steel. 2. Iron and steel bridges. 3. Steel, Structural.
4. Civil engineering. I. West, Adalberto.
TA684 .A53 2023
624.182 1--dc23

For information on all Clanrye International publications
visit our website at www.clanryeinternational.com

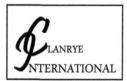

Contents

Preface

Steel is the most widely used component in the construction of buildings and structures. There are many advantages of steel over concrete. Its important properties include recyclability, increased durability, and low maintenance. The design of steel structures typically consists of a two-step analysis and verification procedure. The first step is to evaluate the internal forces and displacements based on the principles of equilibrium and compatibility. In the second step, these internal forces and displacements are compared against corresponding resistance, stiffness and ductility values to ensure structural safety and fitness-for-purpose. Plastic analysis is a method that determines the collapse behavior of structures on the basis of cross sections under proportionally increasing loading. This book examines the analysis and design of steel structures. It is a vital tool for all researching or studying these structures as it gives incredible insights into emerging trends and concepts of civil engineering. Its extensive content provides the readers with a thorough understanding of the subject.

This book is a comprehensive compilation of works of different researchers from varied parts of the world. It includes valuable experiences of the researchers with the sole objective of providing the readers (learners) with a proper knowledge of the concerned field. This book will be beneficial in evoking inspiration and enhancing the knowledge of the interested readers.

In the end, I would like to extend my heartiest thanks to the authors who worked with great determination on their chapters. I also appreciate the publisher's support in the course of the book. I would also like to deeply acknowledge my family who stood by me as a source of inspiration during the project.

Adalberto West

Steel and Limit State Design Method

1.1 Steel: Properties, Advantages and Disadvantages

Properties of Steel

Physical properties of structural steel, as detailed by cl.2.2.4.1 of IS 800:2007, irrespective of its grade may be taken as:

- Unit mass of steel, $p = 7850$ kg/m^3.

- Modulus of elasticity, $E = 2.0 \times 10^5$N/mm^2 (MPa).

- Poisson ratio, $\mu = 0.3$.

- Modulus of rigidity, $G = 0.769 \times 105$ N/mm2 (MPa).

- Coefficient of thermal expansion $c_x = 12 \times 10^{-6}$/°C.

Advantages of Steel

- Steel members have high strength per unit weight. Therefore, a steel member of a small section which has little self-weight is able to resist heavy loads. The high strength of steel results in smaller sections to be used and fewer columns in buildings. The high strength to weight ratio is the most important property for the construction of long span bridges, tall buildings and for buildings on soils with relatively low bearing capacities.

- Steel, being a ductile material, does not fail suddenly, but gives visible evidence of impending failure by large deflections. Also, the ductile nature of the usual structural steels enables them to yield locally at the points of high stress concentrations. The ductility of steel is responsible for relieving the over stressing in certain members by allowing redistribution of stresses due to yielding and thus, preventing premature failures.

- Structural steels are tough, i.e., they have both strength and ductility. Thus, steel members subjected to large deformations during fabrication and erection will not fracture.

- The steel may be bent, hammered, sheared or even the bolt holes may be punched without any visible damage.

- Being light, steel members can be conveniently handled and transported. For this reason, prefabricated members can be frequently provided.

- Properly maintained steel structures have a long life.

- The properties of steel mostly do not change with time. This makes steel the most suitable material for a structure.

- Additions and alterations can be made easily to steel structures.

- They can be erected at a faster rate.

- Steel has the highest scrap value among all building materials. Also, the steel can be reused after a structure is disassembled.

- Steel is the ultimate recyclable material.

Disadvantages of Steel

Steel structures, when placed in exposed conditions, are subjected to corrosion. Therefore, they require frequent painting and maintenance.

The use of weathering steels, however, in suitable applications may eliminate the requirement of frequent painting. Steel structures need fireproof treatment, which increases cost. Furthermore, steel is an excellent heat conductor and, therefore, steel members may transmit enough heat from a burning section or a room of a building to ignite materials with which they are in contact in the adjoining room. Although steel members are incombustible, their strength reduces drastically due to the temperatures reached during fire.

Fatigue of steel is one of the major drawbacks. Fatigue involves a reduction in the strength when steel is subjected to large number of stress reversals and even to a large number of variations of tensile stress.

At the places of stress concentration in the steel sections, under certain conditions, the steel may lose its ductility. Fatigue and very low temperatures aggravate the situation.

1.1.1 Structural Steel

Structural steel has been classified by the Bureau of Indian Standards based on its ultimate or yield strength. For example, Fe 410 steel has minimum tensile strength of 410 N/mm^2. The mechanical properties of steel largely depend on its chemical composition, rolling methods, rolling thickness, heat treatment and stress history.

Properties of Structural Steel

The following are the factor to be considered in mechanical properties of structural steel:

- Chemical composition.

- Rolling method.

- Rolling thickness.

- Heat treatment process.

- Stress history.

The important mechanical properties of steel are ultimate tensile strength, yield or proof stress, ductility, weldability, toughness, corrosion resistance and machinability.

Ultimate Strength or Tensile Strength

Ultimate strength which is the minimum guaranteed ultimate tensile strength (UTS) at which the steel would fail is obtained from a tensile test on a standard specimen, generally called a coupon. A typical specimen as per IS 1608 is shown in figure below. In this test the gauge length Lg and the initial cross-sectional area A_o are important parameters.

The dimensions of the specimens are established to ensure that failure occurs within the designated gauge length. The test coupons are actually cut out from a specified portion of the member for which the tensile strength is required. The initial gauge length is taken as 5.65 A_o in the case of a specimen with a rectangular cross section and five times the diameter in the case of a circular specimen.

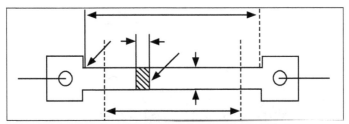

Standard tensile test specimen as per IS 1608.

The coupon is fixed in a tensile testing machine with specified distances between the grips and tested under uniaxial tension. The loads are applied through the threaded ends. A typical stress–strain curve of ordinary and high-strength steel specimen subjected to a gradually increasing tensile load is shown in figure (a) and the stress–strain curve of mild steel specimen is shown in figure (b).

The ultimate tensile strength is the highest stress at which a tensile specimen fails by fracture and is given by,

Ultimate tensile strength = Ultimate tensile load × Original area of cross section.

Note that steel is specified in the code by the characteristic ultimate tensile strength, f_u which is defined as the minimum value of stress below which not more than a specified percentage of corresponding stresses of samples tested are expected to occur. Steel is specified according to the (characteristic) yield strength.

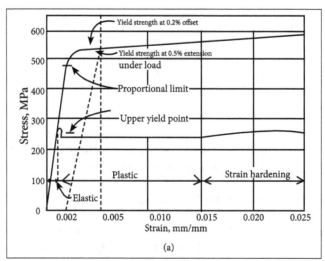

(a)

Stress-strain curves of ordinary and high-strength steel.

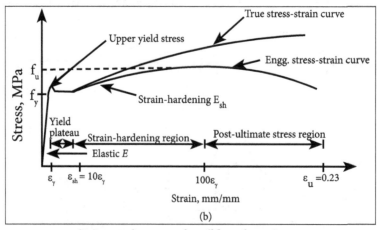

(b)

Stress-strain curve of a mild steel specimen.

Inelastic Cyclic Response

The stress–strain response of most materials under cyclic loading is different from that under single (monotonic) loading. For fatigue analysis, it is necessary to consider the cyclic material behaviour for strength and life calculations.

When steel is subjected to cyclic loading in the inelastic range, the yield plate is suppressed and the stress–strain curve exhibits the Bauschinger effect, in which non-linear response develops at a strain much lower than the yield strain, as shown in following figure. The amplitude of response increases, the stress level for a given strain also increases and can substantially exceed the stress indicated by the monotonic stress–strain curve.

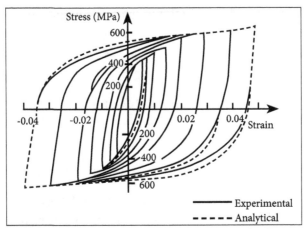

Stress-strain curve of steel subjected to cycle loading.

Characteristic Strength

Variations in material properties should be recognized and taken into consideration in the design process. The material properties that are of greatest importance in the design of structures using steel are yield strength, maximum percentage elongation and Young's modulus. Other properties that are of less importance are hardness, impact resistance and melting point.

If a number of samples are tested for a particular property and the number of specimens with the same strength (frequency) are plotted against the strength then the results approximately fit a normal distribution curve as shown in figure.

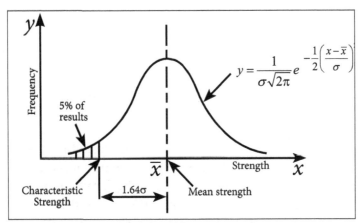

Normal distribution curve.

This curve can be mathematically expressed by the equation shown in figure which can be used to define 'safe' values for design purposes. When defined, this safe value of yield strength is called characteristic strength. If the characteristic strength is defined as the mean strength then from figure, 50% of the material has a characteristic strength below this value and hence is not acceptable. Hence a characteristic value which has a particular chance (often 95%) of being exceeded in any standard tension test is chosen.

Thus the characteristic strength is calculated from the equation,

$$F_k = f_{mean} - 1.64\sigma$$

Where,

σ - Standard derivation for n samples and is given by,

$$\sigma = \left[\frac{\sum(f_{mean} - f)^2}{(n-1)^{0.5}} \right]$$

The characteristic strength of steel is the value obtained from tests at the rolling mills but by the time the steel becomes part of the finished structure its strength might have been reduced. The strength to be used in design calculations is therefore the characteristic strength divided by a partial safety factor.

Ductility

Ductility is also described as the ability of a material to change its shape without fracture. In other words, the ductility of a structure or its members is the capacity to undergo large inelastic deformations without significant loss of strength or stiffness. The stress–strain curve of a material also indicates the ductility. It is the amount of permanent strain, i.e. strain exceeding proportional limit, up to the point of fracture. The ductility of the tension test specimen is measured by determining the percentage elongation. The specified gauge length according to the code is as follows.

$$\text{Gauge length} = 5.65 \sqrt{A_o}$$

Percentage of elongation = Elongated length between gauge point - Gauge length X 100/Gauge length.

The measured elongation is influenced by the gauge length, strain rate of test and failure within or outside the gauge length.

Values of 20% can be obtained for mild steel but are less for high-strength steel. By improper testing one may get percentage elongation values of 20–60 and hence the test houses should be careful with the testing process. A high value is advantageous because it allows the redistribution of stresses at ultimate load and the formation of plastic hinges. For most standard mild steels, the values are greater than the minimum required. However, rerolled steel or improperly controlled steel may give higher strength but less percentage of elongation.

Low Temperature and Toughness (Brittle Fracture)

In structural steel design, toughness is a measure of the ability of steel to resist fracture under impact loading, i.e., the capacity to absorb large amounts of energy. Toughness can be an important design criterion, particularly for structures subject to impact loads (Example: bridges) and for those subject to earthquake loads. Hence both strength and ductility contribute to toughness.

At room temperature, common structural steel is very tough and fails in a ductile manner. At temperatures below 0°C, steel structures sometimes fail suddenly and without warning. A right combination of low temperature, an abrupt change in section size (notch effect) or an imperfection and the presence of tensile stress can initiate a failure called brittle fracture. This may begin as a crack which may propagate and cause the member to fail. Most brittle fractures occur under static load at stress levels which are not excessive, but they may also be due to the dynamic application of a load or some overload.

Structural steels vary greatly in toughness. Highly killed, fine grain steel with a suitable chemical composition or specially heat-treated steel exhibit considerable toughness. IS 2026 and IS 1757 codes allow the use of only those steels that exhibit a minimum energy absorption capacity at a predetermined temperature (Example: 20 J at 23 ± 5°C). In addition to the chemistry of steel, size of plates, residual stress and cold work also affect toughness.

Lamellar Tearing

Lamellar tearing is a form of brittle fracture that may occur in certain welded joints. For example, a tear can occur if a large weld is placed on a thick plate since the shrinkage strains from the welding operation will be large and restrained. The restraint may be developed due to the weld on the far side or due to the member thickness or due to a combination of both the factors.

Generally I-sections are adequately ductile when loaded either parallel or transverse to the rolling direction.

A thin stiffened column is susceptible to lamellar tearing since the flange stiffeners that are welded to the column flange produce a restraint. A large overmatch of electrode and base metal in a full penetration butt weld also tends to increase the possibility of tearing.

The use of fillet welds a joint design that allows weld shrinkage to occur in the rolling direction and the sequence of welding to minimize shrinkage strains are practical methods used to avoid lamellar tearing.

High-Temperature Effects

Steel is not a flammable material. However its strength reduces with rise in temperature. The yields as well as tensile strength at 500°C are about 60–70% of that at room (about 21°C) temperature. The drop in strength is much higher at still higher temperatures.

Hence, steel frames enclosing materials that are flammable require fire protection to control the temperature of steel members for a sufficient time for the occupants to seek safety or for the fire be extinguished before the building collapses. In many cases the building does not collapse even at high temperatures. But the members are deformed beyond acceptable limits and hence have to be replaced.

Hardness

Hardness is a measure of the resistance of the material to indentations and scratching. Several methods are available to determine the hardness of steel and other metals. In all these methods an 'indentor' is forced on to the surface of the specimen. On removal the size of the indentation is measured using a microscope. Based on the size of the indentation, the hardness of the specimen is determined. Brinell hardness (typical value: 150–190) and Vickers hardness tests are used to determine hardness.

Structural Steel Sections

Various kinds of structural steel sections and their technical specifications are as follows:

- Beams

- Channels

- Angles

- Flats

Steel Beams

Steel Beams is taken into account to be a structural part which mainly carries load in flexure meaning bending. Sometimes beams carry vertical gravitational force however are also capable of carrying horizontal loads usually in the case of an earthquake. The mechanism of carrying load in a beam is extremely unique, like the load carried by a beam is transferred to walls, columns or girders that in turn transfer the force to the adjacent structural compression members. The joists rest on the beam in light-weight frame constructions.

Steel beams.

The beams are known by their profile meaning:

- The length of the beam.
- The shape of the cross section.
- The material used.

The most normally found steel beam is that the I beam or the wide flanged beam also known by the name of universal beam or stouter sections as the universal column. Such beams are normally used in the construction of bridges and steel frame buildings.

The most commonly found types of steel beams are varied and they are mentioned below:

- I-Beams.
- Wide flange beams.
- HP shape beams.

Beams expertise tensile, shear and compressive stresses internally because of the loads applied to them. Typically below gravity loads there is a slight reduction within the original length of the beam. This leads to a smaller radius enclosure at the top of the beam thus showing compression. While an equivalent beam at the bottom is slightly stretched enclosing a larger radius are because of tension.

The length of the beam midway and at the bends is the same as it is not under tension or compression and is defined as the neutral axis. The beam is entirely exposed to shear stress above the support. There are some reinforced concrete beams that are fully under compression, these beams are called pre-stressed concrete beams and these are built in such a manner to produce a compression more than the expected tension under the loading conditions.

The pre-stressed concrete steel beams have the manufacturing method like, first the high strength steel tendons are stretched and so the beam is cast over them. Then because the concrete begins to cure, the tendons are released thus the beam is instantly under eccentric axial loads. An internal moment is created due to the eccentric axial load which in turn will increase the moment carrying capacity of the beam. Such beams are generally used in highway and bridges.

Materials Used

In today's modern construction the beams are generally made up of materials like:

- Steel.
- Wood.
- Reinforced concrete.

Steel Channels

Steel channels are used ideally as supports and guide rails. These are roll-formed products. The most metal used for making channels is steel along with aluminum. There are certain variations that are available in the channels category, the categorization is especially on the shape of the channel, the varieties are mentioned below:

- J channels: This type of channel has two legs and a web. One leg is longer. This channel resembles the letter-J.

- Hat channels: This channel has legs that are folded within the outward direction resembling an old fashioned man's hat.

- U channels: This is a most typical and basic channel variety. It has a base called as a web and two equal length legs.

- C channels: In this channel the legs are folded back within the channel and resemble the letter-C. C channels are called as rests.

- Hemmed channels: In this type of channel the top of the leg is folded hence forming double thickness.

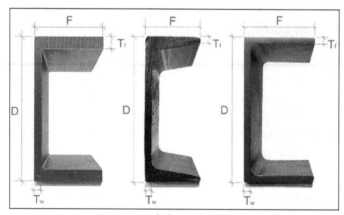

Steel channels.

Application

Steel channels are subjected to a wide array of applications. The application fields are:

- Construction.

- Appliances.

- Transportation.

- Used in making Signposts.

- Used in wood flooring for athletic purposes.

Steel Angle

A steel angle is long steel with reciprocally vertical sides. The steel angles are the most basic kind of roll-formed steel. The most normally found steel angles are formed at a 90° angle and has two legs of equal length. The sides are either equal or of various sizes.

There are certain variations within the steel angles depending on its basic construction. The variations are like if one leg is longer than the other then it is called as L angle.

If the steel angle is something totally different from 90° then it is called as V angle. In some steel angles, double thickness is achieved by folding the legs inward. If the steel angle has same sides then it means that it has identical width. The steel angles are made according to the strength that is needed for the different structures for construction purposes.

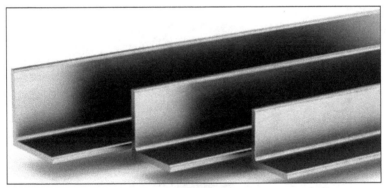

Steel angles.

Applications

The steel angle finds an application in a number of things, they are mentioned below:

- Used in framing.
- Used in trims.
- For reinforcement.
- In brackets.
- Used in transmission towers.
- Bridges.
- Lifting and transporting machinery.
- Reactors.
- Vessels.

- Warehouses.

- Industrial boilers.

- Structural steel angles are used in rolling shutters for fabricating guides for strength and durability.

Steel Flats

Flats are actually thin strips of mild steel having the thickness of the strip normally varying from 10mm to 12mm however thicker flats than this are also available. Steel flats are produced by the use of relatively smooth, cylindrical rolls on rolling mills. Typically the width to thickness ratio of flat rolled products is fairly large. The steel flat bars are manufactured using advanced thickness control technology for controlled thicknesses.

Steel Flats.

The hi-tech machineries enable the production of top grade steel flat bars with superlative flatness and controlled thickness. This product is extremely customized and also the specific sizes according to the client's requirement are produced. After production the flat steels are subjected to a variety of finishes like, painting and galvanizing.

The flat carbon steel is a hot or cold rolled strip product also called as a plate product. These plate products have a size variation between 10mm to 200mm and also the thin flat rolled flat rolled product's size varies from 1mm to 10mm.

Applications

The steel flats are used in a wide array of applications. The varied applications are listed below:

- Railway parts.

- Hand tools.

- Engineering industries.

- Auto components: Two-wheeler, four-wheeler, commercial vehicles.

- Ordinance factories.

- Domestic white goods products.

- Office furniture's.

- Heart pacemakers.

1.1.2 Rolled Steel Sections

Structural steel can be rolled into various shapes and sizes in rolling mills. Usually sections having larger moduli of section in proportion to their cross-sectional areas are preferred. Steel sections are usually designated by their cross-sectional shapes. The shapes of the rolled steel sections available today have been developed to meet structural needs.

Cross-section and size are governed by a number of factors: Arrangement of material for optimum structural efficiency, functional requirements, dimensional and weight capacity of rolling mills and material properties which, for example, inhibit the hot rolling of wide thin elements because of excessive warping or cracking that might occur.

The types of rolled structural steel sections are as follows:

- Rolled steel I-sections.

- Rolled steel channel sections.

- Rolled steel T-sections.

- Rolled steel angle-sections.

- Rolled steel tube-sections.

- Rolled steel bars.

- Rolled steel flats.

- Rolled steel plates.

- Rolled steel sheets.

- Rolled steel strips.

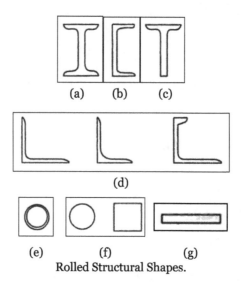

(a)   (b)   (c)

(d)

(e)   (f)   (g)

Rolled Structural Shapes.

Angle sections were probably the first shapes rolled and produced in 1819 in America. I-beam shape was introduced by Zores of France in 1849. By 1870 Channels and tees were developed. All these early shapes were made of wrought iron. The first true skeletal frame structure, the Home Insurance Company Building, was built in Chicago in 1884.

An I-section is designated by its depth and weight, example, I.S.L.B. 500 at 750 N/m means, the I-section is 500 mm deep and the self-weight is 750 N per metre length. A channel section is designated by its depth and weight. For instance, I.S.L.C. 350 at 388 N/m means that the channel section is 350 mm deep and the self-weight is 388 N per metre length.

All standard I-beams and channels have a slope on the inside face of the flange of 16⅔ %. A T-section is designated by its depth and weight, Example., I.S.N.T. 125 at 274 N/m, means the tee section is 125 mm deep and the self-weight is 274 N per metre length.

An angle-section is designated by its leg lengths and thickness. For example, I.S.A. 40 x 25 x 6 mm means, the section has an unequal angle with legs 40 mm and 25 mm in length and thickness of the legs 6 mm.

Steel tubes are designated by their outside diameter and self-weight. Rolled steel bars may be circular or square and are designated by diameter or side respectively, example I.S.SQ. 10 mm means a square bar of 10 mm side and I.S.RO. 10 mm means a round bar of 10 mm diameter.

Steel flats are designated by width and thickness of the section, Example., 30 I.S.F. 10 mm means the flat is 30 mm wide and 10 mm thick. Steel plates are designated by length, width and thickness, example I.S.PL. 2000 mm x 1000 mm x 8 mm, means the plate is 2000 mm long, 1000 mm wide and 8 mm thick.

Steel sheets are designated by length, width and thickness, example, I.S.SH. 2000 mm x 600 mm x 4 mm, means the sheet is 2000 mm long 600 mm wide and 4 mm thick.

Steel strips are designated by width and thickness, Example., I.S.ST. 200 mm x 2 mm, means the strip is 200 mm wide and 2 mm thick.

Choice of Section

The design of steel sections is governed by cross-sectional area and section modulus. It has been seen that a variety of steel sections are rolled, but due to the limitations of rolling mills only a few are available. Also, if a section is in demand, it is rolled regularly but one which is in little demand is rolled only on order and hence costs more. Therefore, the design is not only governed by sectional properties but also on availability of the section in the market, which becomes a major consideration.

1.1.3 Various Types of Loads

The loads are generally classified as vertical loads, horizontal loads and longitudinal loads. The vertical loads consist of dead load, live load and impact load. The horizontal loads comprises of wind load and earthquake load. The longitudinal loads i.e. friction and braking forces are considered in special case of design of bridges, gantry girders etc.

1. Dead Load

Dead loads are permanent or stationary loads that are transferred to structure throughout the life span. Dead load is primarily due to self-weight of structural members, permanent partition walls, fixed permanent equipments and weight of various materials.

2. Imposed Loads or Live Loads

Live loads are either movable or moving loads without any acceleration or impact. They are assumed to be made for the use or occupancy of the building as well as weights of movable partitions or furniture etc. The floor slabs are designed to carry either uniformly distributed loads or concentrated loads whichever produce greater stresses in the part under consideration.

Since it is unlikely that at one specific time all floors will not be simultaneously carrying maximum loading, the code permits some reduction in imposed loads in designing columns, load bearing walls, piers supports and foundations.

3. Impact Loads

Impact load is caused by vibration or impact or acceleration. Thus, impact load is equal to imposed load incremented by some percentage referred to as impact factor or impact allowance depending upon the intensity of impact.

4. Wind Loads

Wind load is primarily horizontal load caused by the movement of air relative to earth.

Wind load is needed to be thought of in design particularly once the heath of the building exceeds two times the dimensions transverse to the exposed wind surface.

The wind load is not crucial because the moment of resistance provided by the continuity of floor system to column connection and walls provided between columns are sufficient to accommodate the result of these forces. Further in limit state method the factor for design load is reduced to 1.2 (DL + LL + WL) once wind is considered as against the factor of 1.5(DL + LL) when wind is not considered. IS 1893 (part 3) code book is to be used for design purpose.

5. Earthquake Load

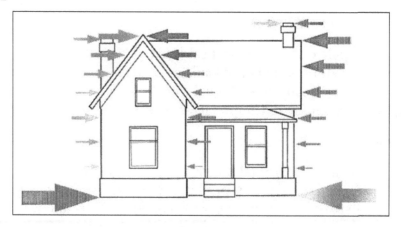

Horizontal earthquake forces (back-and-forth shaking) create 'whipping' forces in all parts of a building. These forces must transfer between parts of the building to the foundation. The above figure shows.

Earthquake loads are horizontal loads caused by the earthquake and will be computed in accordance with IS 1893. For monolithic reinforced concrete structures set within the seismic zone 2 and 3 while not over 5 storey high and importance factor less than 1, the seismic forces are not crucial.

Design Loads: Load Combinations as Per IS Code Recommendations

The following are the various loads to be considered for the purpose of computing stresses, wherever they are applicable:

- Dead load.
- Live load.
- Impact load.
- Longitudinal force.
- Thermal force.

- Wind load.

- Seismic load.

- Racking force.

- Forces due to curvature.

- Forces on parapets.

- Frictional resistance of expansion bearings.

- Erection forces.

Loads are a primary consideration in any building design because they define the nature and magnitude of hazards or external forces that a building must resist to provide reasonable performance (i.e., safety and serviceability) throughout the structure's useful life. The anticipated loads are influenced by a building's intended use (occupancy and function), configuration (size and shape) and location (climate and site conditions).

Ultimately the type and magnitude of design loads affect critical decisions such as material selection, construction details and architectural configuration. Thus to optimize the value (i.e., performance versus economy) of the finished product, it is essential to apply design loads realistically. While the buildings considered in this guide are primarily single-family detached and attached dwellings, the principles and concepts related to building loads also apply to other similar types of construction such as low-rise apartment buildings.

Design Loads for Residential Buildings

Methods for determining design loads are complete yet tailored to typical residential conditions. As with any design function the designer must ultimately understand and approve the loads for a given project as well as the overall design methodology, including all its inherent strengths and weaknesses.

Since building codes tend to vary in their treatment of design loads the designer should as a matter of due diligence identify variances from both local accepted practice and the applicable building code relative to design loads as presented even though the variances may be considered technically sound. Complete design of a home typically requires the evaluation of several different types of materials.

Some material specifications use the allowable stress design (ASD) approach while others use load and resistance factor design (LRFD). Therefore for a single project it may be necessary to determine loads in accordance with both design formats.

The determination of individual nominal loads is essentially unaffected. Special loads such as flood loads, ice loads and rain loads are not addressed herein. The reader is referred to the ASCE 7 standard and applicable building code provisions regarding special loads.

Load Combinations

Keeping the aspect specified in 8.0, the various loads should therefore be combined in accordance with the stipulations in the relevant design codes. In the absence of such recommendations, the following loading combinations, whichever combination produces the most unfavorable effect in the building, foundation or structural member concerned may be adopted (as a general guidance). It should also be recognized in load combinations that the simultaneous occurrence of maximum values of wind, earthquake, imposed and snow loads is not likely:

- DL

- DL + IL

- DL + WL

- DL + EL

- DL + TL

- DL + IL + WL

- DL + IL + EL

- DL + IL + TL

- DL + WL + TL

- DL + EL + TL

- DL + IL + WL + TL

- DL + IL + EL + TL

(DL = dead load, IL = imposed load, WL = wind load, EL = earthquake load, TL = temperature load). When snow load is present on roofs, replace imposed load by snow load for the purpose of above load combinations.

The relevant design codes shall be followed for permissible stresses when the structure is designed by working stress method and for partial safety factors when the structure is designed by limit state design method for each of the above load combinations.

Whenever imposed load (IL) is combined with earthquake load (EL), the appropriate part of imposed load as specified in IS : 1893-1984 should be used both for evaluating earthquake effect and for combined load effects used in such combination.

For the purpose of stability of the structure as a whole against overturning, the restoring moment shall be not less than 1. 2 times the maximum overturning moment due to dead load plus 1. 4 times the maximum overturning moment due to imposed loads.

In cases where dead load provides the restoring moment, only 0.9 times the dead load shall be considered.

The restoring moments due to imposed loads shall be ignored. In case of high water table the effects of buoyancy have to be suitably taken into consideration. In case of high water table the factor of safety of 1.2 against uplift alone shall be provided.

The structure shall have a factor against sliding of not less than 1·4 under the most adverse combination of the applied loads and forces. In this case only 0·9 times the dead load shall be taken into account.

Where the bearing pressure on soil due to wind alone is less than 25% of that due to dead load and imposed load, it may be neglected in design. Where this exceeds 25 percent foundation may be so proportioned that the pressure due to combined effect of dead load, imposed load and wind load does not exceed the allowable bearing pressure by more than 25%. When earthquake effect is included the permissible increase is allowable bearing pressure in the soil shall be in accordance with IS : 1893-1984.

Reduced imposed load (IL) specified in Part 2 of this standard for the design of supporting structures should not be applied in combination with earthquake forces. Other loads and accidental load combinations not included should be dealt with appropriately.

Crane load combinations are covered under/Part 2 of this standard:

S. No.	Load combination	Loads
1	Stresses due to normal loads	Dead load, live load, impact load and centrifugal force.
2	Stresses due to normal loads + occasional loads.	Normal load as in (1) + wind load other lateral loads. longitudinal forces and temperature stresses.
3	Stresses due to loads during erection	-
4	Stresses due to normal loads + occasional loads + Extra-ordinary loads wind like seismic excluding wind load.	Loads as in (2) + with seismic load instead of wind.

1.1.4 Design Philosophies

Design of steel structures consists of the design of steel members and their connections so that they will safely and economically resist and transfer the applied loads. Although the ease of fabrication and feasibility of connections of various members may ultimately dictate the choice, the process begins with the selection of a section with minimum weight per unit length.

Once the trial section is selected, the designer checks its safety. This is where different approaches to design come into play. The design of structural steel elements is based on one or more of the following criteria:

- Attainment of initial yielding.

- Attainment of full yielding.

- Tensile strength.

- Critical buckling.

- Maximum deflection permitted.

- Stress concentration.

- Fatigue.

- Brittle fracture.

A steel structure may be designed by any one of the three design philosophies: Elastic or working stress method, Plastic or limit design method and Limit State method.

Working Stress Method

Working stress method is the elastic method of design. The worst combination of working (service) loads is ascertained and the members are proportioned on the basis of working stresses. These stresses should never exceed the permissible stresses as laid down by the code.

The permissible stresses are some fraction of the yield stress of the material and may be defined as the ratio of the yield stress to the factor of safety. It may also be defined as the ratio of strength of the member to the expected force. In other terms, when the yield point is well defined, the factor of safety is defined as the ratio of the yield stress to the maximum expected stress.

The concept of introducing a factor of safety is to make the structure safe. It accounts for the following:

- The analysis methods are based on assumptions and do not give the exact stresses.

- Structural members may temporarily be overloaded under certain circumstances.

- The stresses due to fabrication and erection are not considered in the design of ordinary structures.

- The secondary stresses may be appreciable.

- Underestimation of the future live loads.

- Stress concentrations.

- Unpredictable natural calamities.

Plastic Method

Steel is a ductile material and from the stress-strain curve it is observed that higher loads than in the elastic method can be applied over the structure. This is due to the fact that a major portion of the curve lies beyond the elastic limit. This extra strength is termed reserve strength and forms the basis of plastic design method. This is an aspect of limit design, which confines the structural usefulness up to the plastic strength or ultimate load carrying capacity.

This method is based on failure conditions rather than working load conditions. In this method of design failure implies collapse or extremely large deformations, thus the structure fails at a much higher load, called the collapse load, than the working load. In the plastic design method, the working loads are multiplied by the load factor and the cross sections of members are selected and designed on the basis of the collapse strength.

The term plastic is used because, at failures, parts of the member will be subjected to very large strains large enough to put the member into plastic range. When the entire cross section becomes plastic, infinite rotation takes place and a plastic hinge is formed. When sufficient plastic hinges are formed in the structural member at the maximum stressed locations, a collapse mechanism is formed. Since the actual loads will be less than the collapse loads by a factor of safety (load factor), the members designed will be safe.

Limit State Method

Limit state method is similar to plastic design which considers most critical limit states of strength and serviceability. The acceptable limit for the safety and serviceability requirements before failure occurs is called a limit state. The objective of design is to achieve a structure that will not become unfit for use with an acceptable target reliability.

In other words, the probability of a limit state being reached during its lifetime should be very low. In general, the structure should be designed on the basis of the most critical limit state and should be checked for other limit states. A structure should be designed to withstand safely all loads likely to act on it.

For ensuring the design objectives, the design should be based on characteristic values for material strengths and applied loads (actions), which take into account the probability of variations in the material strength and in the loads to be supported. The characteristic value should be based on statistical data. Where such data is not available, they should be based on experience.

The design values are derived from the characteristic values through the use of partial safety factors, both for material strengths and for loads. The reliability of design is ensured by requiring that load factors are applied to the service loads and then theoretical strength of the member is reduced by the application of a resistance factor. The criterion to be satisfied in the selection of a member is,

Design action ≤ Design strength.

$$\text{Design action} = Q_d = \sum_k \gamma_{fk} \, Q_{ck}$$

Where,

Q_{ck} = Characteristic actions that are not expected to be exceeded with 5% probability during the life of the structure which include self-weight, live load or imposed load, crane load, wind load, earthquake load.

γ_{fk} = Partial safety factor for different loads k.

The possibility of inaccurate assessment of the load and the uncertainty in the assessment of effects of the load and the uncertainty in the assessment of the limit states being considered.

For three types of loads (k = 1, 2, 3), viz., dead load, live load/crane load and wind load/ earthquake load, Q, may be written as:

$$Q_d = \gamma_{f1} \, Q_{c1} + \gamma_{f2} \, Q_{c2} + \gamma_{c2} + \gamma_{f3} \, Q_{c3}$$

Where,

γ_{f1} = Partial safety factor for dead load

γ_{f2} = Partial safety factor for live load/crane load/erection load

γ_{f3} = Partial safety factor for wind load/earthquake load

Q_{c1} = Characteristic action for dead load

Q_{c2} = Characteristic action for live load/crane load/erection load (1, 3 = Characteristic action for wind load/earthquake load).

The section designed should also satisfy the serviceability requirements, such as limitations of deflection and vibration and should not collapse under accidental loads such as from explosions or impact or due to consequences of human error to an extent not originally expected to occur.

1.2 Limit State Design Method

The limit state method of design was developed to take account of all conditions that can make the structure unfit for use, considering actual behaviour of materials and structures. IS 800:2007, the relevant code of practice, applicable to the structural use of hot-rolled steel is largely based on limit state method of design. Since, it still retains the working stress method which was in use for last several decades.

The code recommends the working stress method in situations where limit state method cannot be adopted conveniently and confidently. Both the design philosophies have therefore been incorporated in the body of the text, but with emphasis on limit state method of design being more realistic and resulting in economical designs. The limit state method of design also known as load and resistance factor method.

Since 1974, it has been in use in Canada and Europe as Limit States design and as Load and Resistance Factor method in America. There are basically two categories of limit states, strength and serviceability. The acceptable limit for the safety and serviceability requirements before failure occurs is called a limit state. Strength limit states are based on the load carrying capacity of structures and include plastic strength, buckling, fracture, fatigue, overturning, etc.

The serviceability limit states refer to the performance of the structures under service loads and include deflections, vibrations, deteriorations, corrosion, ponding, etc. The specifications in the code concentrate more on the strength limit state and give some freedom of judgment to the designer regarding the serviceability area. It is so because the safety of the user is most important but in no way serviceability should be compromised.

In limit state design, basically statistical methods have been used for determination of loads and material properties with a small probability of structure reaching the limit states of strength and serviceability. This implies recognition of the fact that loads and material strength vary, approximations are used in design and imperfections in fabrication and erection affect the strength in service.

All these factors can only be realistically assessed in statistical terms. However, it is not yet possible to adopt a complete probability basis for design and therefore the method adopted ensures safety by using suitable factors. Partial factors of safety are introduced to take account of all the uncertainties in loads, material strength, etc.

Limit States for Steel Design

Limit states are the states beyond which the structure no longer satisfies the specified performance requirements. All the relevant limit states as follows should be considered, but usually it will be appropriate to design on the basis of strength and stability at the ultimate loads and then checking for deflection under serviceability loading. The recommendations regarding the other limit states should also be considered when appropriate.

1.2.1 Limit States of Strength and Serviceability

Limit States of Strength are those associated with failures (or imminent failure), under the action of probable and most unfavourable combination of loads on the structure using the appropriate partial safety factors, which may endanger the safety of life and property.

Limit state of strength is also called ultimate limit state and includes:

- Loss of equilibrium of the structure as a whole or any of its parts or components.
- Loss of stability of the structure (including the effect of sway where appropriate and overturning) or any of its parts including supports and foundations.
- Failure by excessive deformation and formation of mechanism, general yielding, rupture of the structure or any of its parts or components or buckling.
- Fracture due to fatigue.
- Brittle fracture.

Limit States of Serviceability

Limit States of Serviceability are limit states beyond which specified service criteria are no longer met. These include the following:

- Deformation and deflections, which may adversely affect the appearance or effective use of the structure or may cause improper functioning of equipment or services or may cause damages to finishes and non-structural members.
- Vibrations in the structure or any of its components causing discomfort to people, damages to the structure or its contents which may limit its functional effectiveness. Special consideration should be given to floor systems susceptible to vibration, such as large open-floor areas free of partitions to ensure that such vibrations are acceptable for the intended use and occupancy.
- Repairable damage due to fatigue (due to wind-induced oscillations).
- Corrosion and durability.
- Ponding of structures.

Limit State Design Concepts

Steel structures are vital during a variety of land-based applications, including industrial, infrastructural and residential sector. The fundamental strength members in steel structures include support members, plates, stiffened panels/grillages and box girders. Throughout their lifetime, the structures constructed using these members are subjected to various types of loading that is for the most part operational but may in some cases be extreme or even accidental.

Steel-plated structures are probably be subjected to various types of loads and deformations arising from service requirements that may range from the routine to the extreme or accidental. The mission of the structural designer is to design a structure which will withstand such demands throughout its expected lifetime.

The structural design criteria used for the utility Limit State Design of steel-plated structures are commonly based on the limits of deflections or vibration for normal use. In reality, excessive deformation of a structure can also be indicative of excessive vibration or noise and so, certain interrelationships can exist among the design criteria being defined and used for separate convenience.

The SLS criteria are commonly defined by the operator of a structure or by established practice, the first aim being efficient and economical in-service performance without excessive routine maintenance or down-time. The suitable limits essentially depend on the type, mission and arrangement of structures.

The structural design criteria to stop the ultimate limit state design are based on plastic collapse or ultimate strength. The simplified ULS design of the many kinds of structures has in the past tend to rely on estimates of the buckling strength of components, typically from their elastic buckling strength adjusted by a simple plasticity correction. This is represented by point A in figure. In such a design scheme based on strength at point A, the structural designer does not use detailed information on the post-buckling behavior of component members and their interactions.

The true ultimate strength represented by point B in figure is also higher although one will never be sure of this since the actual ultimate strength is not being directly evaluated. In any event, as long as the strength level associated with point B remains unknown, it is difficult to determine the real safety margin. Hence, more recently, the design of structures such as offshore platforms and land-based structures such as steel bridges tend to be based on the ultimate strength.

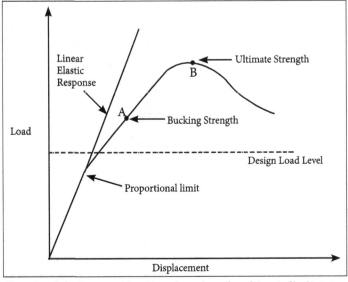

Structural design considerations based on the ultimate limit state.

The safety margin of structures can be computed by a comparison of ultimate strength with the extreme applied loads (load effects) as depicted in following figure. To obtain a safe and economic structure, the ultimate load-carrying capacity as well as the design load should be assessed accurately. The structural designer may even desire to estimate the ultimate strength not only for the intact structure, but also for structures with existing or premised damage, in order to assess their damage tolerance and survivability.

Ultimate Limit State Design of Steel Structures reviews and describes both fundamentals and practical design procedures in this field. Designs should ensure that the structure does not become unfit and unserviceable for the use for that it is intended to. The state at which the unfitness occurs is known as a limit state.

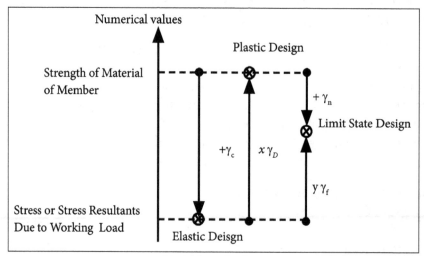

Limit-State Design.

Above figure shown how limit-state design shows separate factors γ_f, that reflects the combination of variability of loading γ_l, material strength γ_m and structural performance γ_p. In the elastic design approach, the design stress is achieved by scaling down the strength of material or member using a factor of safety γ_e as indicated in figure, while the plastic design compares actual structural member stresses with the effects of factored-up loading by using a load factor of γ_p.

Special features of limit state design method are:

- Serviceability and the ultimate limit state design of steel structural systems and their components.

- Due importance has been provided to all probable and possible design conditions that could cause failure or make the structure unfit for its intended use.

- The basis for design is entirely dependent on actual behaviour of materials in structures and the performance of real structures, established by tests and long-term observations.

- The main intention is to adopt probability theory and related statistical methods in the design.

- It is possible to take into account a number of limit states depending upon the particular instance.

- This method is more general in comparison to the working stress method. In this method different safety factors can be applied to different limit states, which is more rational and practical than applying one common factor (load factor) as in the plastic design method.

- This concept of design is appropriate for the design of structures since any development in the knowledge base for the structural behaviour, loading and materials can be readily implemented.

1.3 Probabilistic Basis for Design Riveted, Bolted and Pinned Connections

Probabilistic Basis for Limit State Design

The safety of structure is of prime importance. Safety margins in the form of factor of safety in working stress design approach and load factors in plastic design approach are used to ensure safety against the risk of failure. Although, in general, the structures designed based on these design philosophies were found to be safe but lacked in scientific justifications underlying the provision of safety margins.

The main parameters in analysis and design of the loads are the material properties and the dimensions. The statistical variation of these design parameters is usually ignored in conventional practice and representative unique values such as minimum guaranteed values, limit loads or ultimate loads are used. The conventional approach in design practice may be compared to a kind of worst case analysis.

The maxima of loading and the minima of strength are treated not only as representative of design situations, but also of simultaneous occurrence. This is the basis on which unknown parameters are computed. Actually, magnitude and frequency relationship for both load and strength must be considered to avoid unrealistic results. Therefore, any realistic, rational and qualitative representation of safety must be based on statistical and probabilistic analysis. Both the partial safety factors for loads and materials specified by IS: 800-2007 for the limit state design are based on probabilistic concepts.

Consideration of physical systems, in general, leads to the observation that the magnitude of almost every measurable parameter tends to vary in a random fashion. Variables are characterized by spectra of values rather than by unique values. The observation appears to be true for a large class of parameters, including a strength property of a material, a load magnitude, area of a structural section, ultimate resistance of a structure, etc.

When a typical population of values is plotted on a graph as magnitude versus frequency, such a plot generally tends toward a stable, predictable distribution as the size of the sample is increased. A variable displaying such a characteristic is called a random variable or variate (figure (a)).

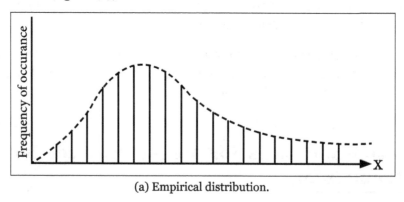

(a) Empirical distribution.

Experimental data can be represented in the form of a histogram or bar graph, as shown in the figure (b),with the abscissa representing sample values or events and the ordinate representing either the number of samples having a certain value or the frequency of occurrence of a certain value. Each bar represents a single sample value or a range of values.

If the ordinate is the percentage of values rather than the actual number of values, the graph is referred to as a relative frequency distribution. In such a case, the sum of the ordinates will be 100%. If the abscissa values are random events and enough samples are used, each ordinate can be interpreted as the probability, expressed as a percentage, of that sample value or event occurring.

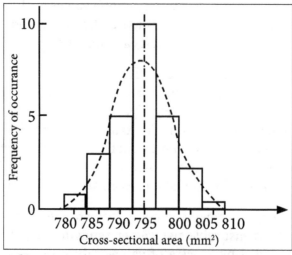

(b) Variation in cross-sectional area of steel section.

If the actual frequency distribution is replaced by a theoretical continuous function that closely approximates the data, it is called a probability density function. Such a function is illustrated in figure(c). Probability functions are designed so that the total area under the curve is unity. That is, for a function $f(x)$,

$$\int_{-\infty}^{+\infty} f(x)dx = 1$$

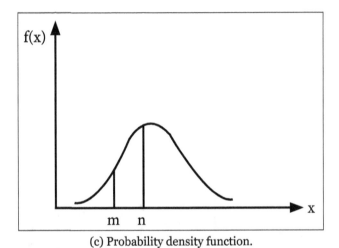

(c) Probability density function.

Which means that the probability that one of the sample values or events will occur is 1.0. The probability of one of the events between m and n in figure(c) equals the area under the curve between m and n,

$$\int_{m}^{n} f(x)dx$$

A typical plot of values usually approximates some known distribution type such as the normal, gamma, log-normal or beta. Thus, most engineering variables may be realistically described by defining statistics, such as mean values and standard deviations. For example, the cross-sectional area of a tension member may be described by a mean value μ and a measure of variation about the mean, called the standard deviation σ.

The representation (σ, μ) is known as moment or couple. This forms the basis for the Level II reliability method also known as first-order-second moment reliability method for structural safety. For safety of structure, IS:800-2007 makes use of partial safety factors, based on probabilistic approach with approximations.

Joint Efficiency

A numerical value expressed as the ratio of the strength of a riveted, welded or brazed joint to the strength of the parent metal.

Riveted Connections

Tearing of the plate: If the force is just too massive the plate could fail in tension on the row. the utmost force allowed during this case is given by,

$$P_1 = s_t (p-d)t$$

Where,

s_t = Allowable tensile stress of the plate material.

p = Pitch.

d = Diameter of the rivet hole.

t = Thickness of the plate.

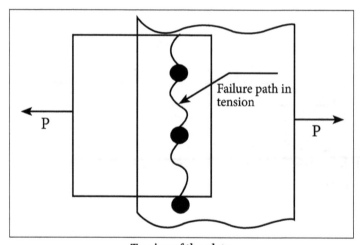

Tearing of the plate

Efficiency of the one riveted joint are often obtained as quantitative relation between the utmost of and also the load carried by a solid plate that is $s_t p_t$. so efficiency (η) = min $s_t p_t$. In a double or triple riveted joint the failure mechanisms are more then those mentioned above. The failure of plate on the outer row could occur within the same means as above but additionally the inner rows could fail.

As an example in an exceedingly double riveted joint the plate could fail on the second row however so as to try and do that ,the rivets within the initial row should fail either by shear or by crushing. So the utmost allowable load such the plate doesn't tear within the second row is,

$$P_4 = s_t (p-d)t + \min(P_2, P_3).$$

Further the joint could fail by,

- Cutting of rivets in each rows.

- Crushing of rivets in each rows.

- Cutting of rivet in one row and crushing within the different row.

Pitch of the rivet: The centre to centre distance of two adjacent rivets in a row is called pitch.

Staggered Pitch: The distance between any two consecutive rivets in a zig-zag riveting, measured parallel to the direction of stress in the member is called staggered pitch.

Reverting: Reverting is a method of joining together Structural Steel Components by inserting ductile metal pins, called rivets into holes of the components to be connected from departing.

Simple Connections: Riveted, Bolted and Pinned Connections

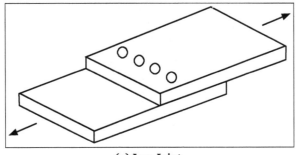

(a) Lap Joint.

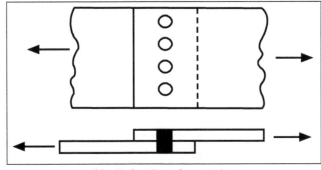

(b) Single Riveted Lap Joint.

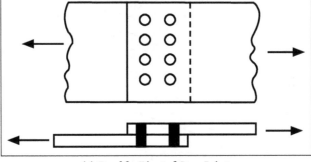

(c) Double Riveted Lap Joint.

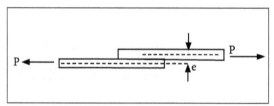

(d) Eccentricity in Lap Joint.

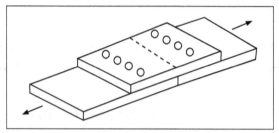

(e) Single Cover Butt Joint.

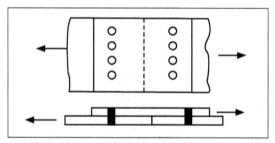

(f) Single Riveted Single, Cover Butt Joint.

Types of Rivet Joints

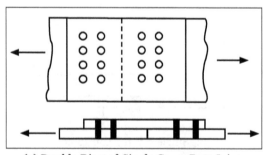

(g) Double Riveted Single Cover Butt Joint.

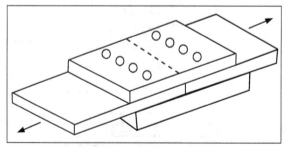

(h) Double Cover Butt Joint.

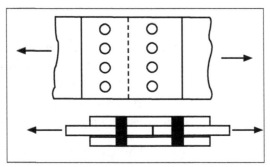

(i) Single Riveted, Double Cover Butt Joint.

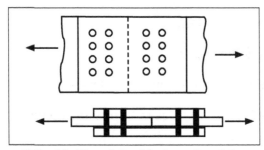

(j) Double Riveted, Double Cover Butt Joint.

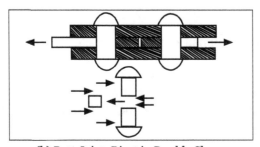

(k) Butt Joint, Rivet in Double Shear.

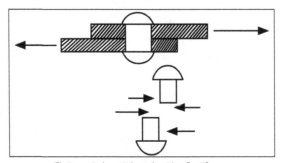

(l) Lap Joint, Rivet in Single Shear.

Strength and Capability of Rivet

Strength of a riveted joint is evaluated taking all attainable failure methods within the joint into consideration. Since, rivets square measure organized in an exceedingly periodic manner, the strength of joint is sometimes calculated considering one pitch length of the plate. There are four possible ways a single rivet joint may fail.

Pin Connections

When two structural members are connected by means of a cylindrical shaped pin, the connection is called a pin connection. Pins are manufactured from mild steel bars with diameters ranging from 9 to 330mm. Pin connections are provided when hinged joints are required, i.e., for the connection where zero moment or free rotation is desired. Introduction of a hinge simplifies the analysis by reducing indeterminacy.

These also reduce the secondary stresses. These connections cannot resist longitudinal tension. For satisfactory working it is necessary to minimize the friction between the pin and members connected. High grade machining is done to make the pin and pin hole surface smooth and frictionless.

Pins are provided in the following cases:

- Tie rod connections in water tanks and elevated bins.

- As diagonal bracing connections in beams and columns.

- Truss bridge girders.

- Hinged arches.

- Chain-link cables suspension bridges.

Various types of pins used for making the connections are forged steel pin, undrilled pin and drilled pin. To make a pin connection, one end of the bar is forged like a fork and a hole is drilled in this portion. The end of the other bar to be connected is also forged and an eye is made.

A hole is drilled into it in such a way that it matches with the hole on the fork end bar. The eye bar is inserted in the jaws of the fork end and a pin is placed. Both the forged ends are made octagonal for a good grip. The pin in the joint is secured by means of a cotter pin or screw, as shown in below figure.

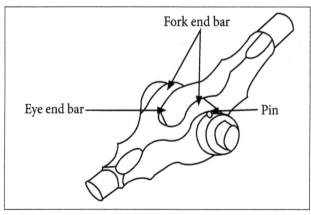

Pin Connection.

Specifications

Shear

Structurally a pin may be treated as a cylindrically shaped beam. Assuming that bending does not exceed yield strength, the shear stress is maximum at the neutral axis of the pin and may be computed by beam-shear formula.

$$\tau_{vf} = \frac{V\overline{AY}}{Id}$$

or,

$$\tau_{vf} = \frac{V\dfrac{\pi d^2}{8}\dfrac{2d}{3\pi}}{\dfrac{\pi d^4}{64}d}$$

or,

$$\tau_{vf} = \frac{16V}{3\pi d^2} \qquad\qquad ...(1)$$

$$\tau_{vf, avg} = \frac{3}{4}\tau_{vf}$$

where ζ_{vf} is the maximum shear stress in pin (100 MPa), V is the shear force at the section in Newtons, d is the diameter of the pin in mm, $\overline{AY}$ is the moment of the area of the cross-section above the section under consideration about the neutral axis and I is the moment of inertia of cross-section.

Calculations of shear stress from the beam shear formula give considerable error when the span-to-depth ratio is small, as it usually is for pins. This necessitates the use of nominal shear stress based on the uniform stress distribution over the pin section, in which case given by

$$\zeta_{vf. Avg} = \zeta_{vf} = V/A$$

The allowable shear stress in the pins is adopted the same as that for power driven rivets.

Bearing: A pin in bearing is designed in a similar manner as for rivet. A uniform bearing stress may be assumed for a proper fit between the plates and the pin,

$$\text{Bearing force} = dt\,\sigma_{pf} \qquad\qquad ...(2)$$

Where,

d - Diameter of the pin in mm,

t - Thickness of bar in mm,

σ_{pf} - Permissible bearing stress (300 MPa).

Flexure

Flexure is most critical in case of pins. The members joined by pin connections are separated by some distance because of the following reasons:

- To prevent friction.

- To allow for rivet heads, if the member is built up.

- To facilitate painting.

Large bending moments are generated due to these reasons and the pin diameter is, therefore, generally governed by flexure.

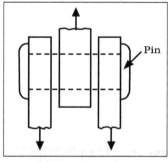

(a) Pin connection.

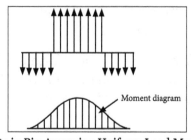

(b) Moments in Pin Assuming Uniform Load Moment in Pin.

Figure (a) shows a pin connection. For a given bending moment M (figure (b)), assuming the stress to be within the elastic limit, the maximum fibre stress can be obtained by the equation,

$$M = \sigma_b Z$$

or,

$$= \sigma \frac{}{32}$$

$$d = \left(\frac{32M}{\pi \sigma_b} \right)^{1/3}$$
...(3)

Where,

M-Bending moment in N mm,

σ_b-Bending stress ($0.66 f_y$).

Design

The design of a pin is similar to that of a rivet. A pin may be assumed to be a large rivet subjected to shear, bearing and flexure. Rivets, as we have seen are critical in shear and bearing. On the other hand, pins are critical in flexure. Therefore, an efficient design can be obtained by determining the pin size on the basis of its bending strength.

- The forces are calculated and points on the pin at which the forces from the members are transferred are ascertained.

- Maximum bending moment on the pin is calculated.

- The diameter of the pin is calculated by equation (3).

- The average shear stress is calculated. It should be less than the allowable shear stress of 100 MPa.

- The bearing stress is calculated. It should be less than the allowable bearing stress of 300 MPa.

Problems

1. Let us design a plate in which the axial load tension 520 kN to be provided with a splice joint. The size of the plate is 400 mm × 20 mm.

Solution:

Given:

Size of plate = 400 mm × 20 mm

Force P = 520 kN

Permissible shear stress = 80 N/mm²

Permissible bearing stress = 250 N/mm²

Permissible tensile strength of plate = 150 N/mm²

Shear strength of one 20 mm ø rivet $= 2\dfrac{\pi d^2}{4}$ is,

$$= \frac{2(\pi)(21.5)^2}{4}(20)$$

$$= 58058 \text{ N} \hspace{6cm} ...(1)$$

Permissible bearing strength of one rivet $= dt\, b_b$

$$= 21.5\,(20)\,(250)$$

$$= 107500 \text{ N} \hspace{5.5cm} ...(2)$$

$\therefore$ Strength of one rivet = Least value of shearing and bearing strength of one rivet

$$= 58058 \text{ N}$$

$\therefore$ No. of rivets requires $= \dfrac{520(1000)}{58058} = 8.95 \approx 9$ rivets.

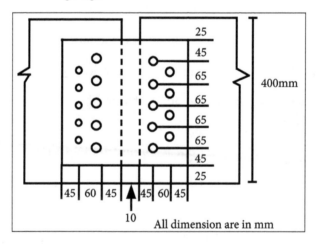

Thickness of Cover Plate Required

Let the width of cover plate = 400 mm

$$= 400 - 50 - 2(21.5)$$

$$= 242.5 \text{ mm}$$

Net area of cross section of each cover plate $= \dfrac{p}{2(f_t)}$

$$= \dfrac{520000}{2(150)}$$

$$= 1733.33 \text{ mm}^2$$

Thickness of cover plate $= \dfrac{1733.33}{242.5}$

= 7.14 mm

= 45 + 60 + 45 + 10 + 45 + 60 + 45

= 310 mm.

∴ Provide a double rivet double cover joint rivet plate size 310 × 242.5 × 8 mm with rivet rows.

2. Let us design a double riveted double cover butt joint connecting 2 plates of 12 mm thick. Adopt power driven rivets take f_y = 250 MPa. Let us calculate the efficiency of the joint.

Solution:

Given:

$f_y = 250 MPa$

Thick = 12 mm

Shearing stress of one rivet = 100N/mm²

Bearing stress of one rivet = 300 N/mm²

∴ Shear strength of rivet bitch length in double rivet cover joint,

$$= 2\left[2\left(-\!\!-\!\!-\right)f\right]$$

Rivet Cover Joint,

$$= 2\left[2\left(\frac{3.4\times(23.5)^2}{4}\right)100\right]$$

= 173407 N

= 2 (d_z f_b)

= 2 (23.5 (12) 300

= 169200 N

Rivet value = 169200 N

Fearing Strength Pitch length = (p – d) t f_t

= (p – 23.5) (12) (250)

$(p - 23.5)\,(12)\,(250) = 169200$

$P = 56.4 + 23.5$

$= 79.9$ say 80 mm

Strength of solid plate $= Pt\,f_t$

$= f_o\,(12)\,(250)$

$= 240000$ N

$\therefore$ Efficiency of solid joint $= \dfrac{169200}{240000} \times 100 = 70.55\%$

3. Let us design a single riveted double cover butt joint used to connect two plate 12 mm thick. The rivets are used as power driven 18 mm in diameter at pitch of 60 mm.

Solution:

Given:

Thick = 12mm

Pitch = 60 mm

Since it is power driven rivets, let us assume the following permissible stress values,

Axial tensile stress = 100 N/mm²

Shear stress = 100 N/mm²

Bearing stress = 300 N/mm²

Since the joint is double cover butt joint the rivets are in double shear. Since, the joint is a single riveted joint. The number of rivets covered per pitch length = 1 finish diameter of the revert hole = 18 + 1.5 = 19.5 mm.

Considering on pitch length of the joint:

(a) Tearing strength per pitch length:

$P_t = (p-d)t\,f_t$

$\quad = (60 - 19.5)\,(12)\,(100)$

$\quad = 48600$N

(b) Shearing strength per pitch length:

$= (1)\,(2)\left(\dfrac{\pi d^2}{4} \cdot fs\right)$

$$= (1)(2)\left[\frac{\pi(19.5)^2}{4}(100)\right] = 59699 \text{ N}$$

(c) Bearing strength per pitch length

$$= (1) \text{ dt } f_b$$

$$= (1)(19.5)(12)(300)$$

$$= 70200 \text{ N}$$

∴ Safe load per pitch length = 48600 N

Strength of solid plate per pitch length $= p_t f_t$

$$= 60(12)(100)$$

$$= 72000 \text{ N}$$

∴ Efficiency of the joint $= \dfrac{48600}{72000} \times 100$

$$= 67.5\%$$

4. The main plates are 12 mm thick connected by 18 mm diameter rivets at a pitch of 100 mm. Let us design the cover plates and also find the percentage reduction in the efficiency of the point if the plates are lap jointed.

Solution:

Given:

Thick = 12 mm

Rivets diameter = 18mm

Pitch =100 mm

Since the joint is a butt joint with two cover plates, the rivets are in double shear. Since the joints is single riveted joint, the member of rivets covered per pitch length = 1. Finished diameter of rivet, d = 22 + 1.5 = 23.5 mm.

Consider One Pitch Length of the Joint

(a) Tearing strength per pitch length:

$$= P_t = (P - d)t/t$$

$$= (100 - 23.5)\ 16 \times 150 = 1,83,600 \text{ N}$$

(b) Shearing strength per pitch length:

$$= [1] \times 2f_s \frac{\pi d^2}{f} = 1 \times 2 \times 100 \times \frac{(2.3)^2}{4} = 86,740N$$

(c) Bearing strength per pitch length:

$$= [1] \times f_b \, dt = 1 \times 300 \times 23.5 \times 16$$

$$= 1,12,800 \text{ N}$$

Safe load per. pitch length = 86,740 N

Strength of the solid plate per pitch length = $p_t f_t$ = 100 × 16 × 150 = 2,40,000 N

Efficiency of the joint $= \dfrac{86,740}{2,40,000} \times 100 = 36.14\%$

Result

Safe load per pitch length = 86,740 N

Efficiency of the joint = 36.14%

Bolted Connection

Bolts can be used in both bearing-type connections and slip-critical connections. Slip-critical connections are used for structures designed for vibratory or dynamic loads such as bridges, industrial buildings and buildings in regions of high seismicity.

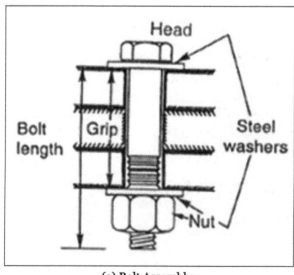

(a) Bolt Assembly.

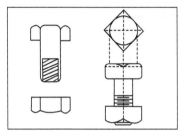

(b) Ordinary Hexagonal Head bolt or ordinary Square head bolt.

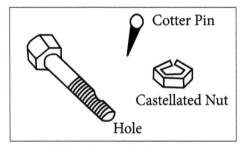

(c) Locking nut by Cotter pin.

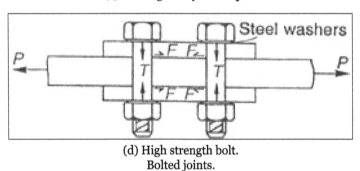

(d) High strength bolt.
Bolted joints.

Efficiency of a Bolted Connection

$$\text{Efficiency}[\eta] = \frac{\text{Strength of joint}}{\text{Strength of solid plate}} \times 100$$

Prying Action

Moment resisting beam to column connection often contain regions in which the bolts will be required to transfer load by direct tension, such as the upper bolt in end plate connection. We should consider an additional force induced in the bolts as a result of so called prying action.

Uses

- Bolts can be used for making end connections in tensions and compression member.

- Bolts can also be used to hold down column bases in position.

- They can be used as separators for purlins and beams in foundations, etc.

Comparison between Gauges distance and pitch of the bolt:

Gauge Distance	Pitch of the Bolt
It is the distance between the two consecutive bolt of adjacent rows.	It is the center spacing of the bolts in a row.
It is measured at right angles to the direction of load.	It is measured along the direction of load.

High Tension Bolts

High tension bolts develop a high initial tension which facilitates clamping or gripping of the components converted.

Spitting of Plates

The failure mode of two connected plate due to tearing or shearing is called spitting of plate.

Type of Bolts

Some type of bolts are:

- Unfinished bolts or black bolts or 'C' grade bolts.

- Turned bolts such as:

  ○ Precision bolts or 'A' grade bolts.

  ○ Semi-precision or 'B' grade bolts.

- Rubbed bolts or fluted bolts.

- High strength friction grip bolts.

Problems

1. The doubly bolded lap joint for plates 16 mm thick that carry its full load. Take permissible axial tension in plate 150 N/mm².

Solution:

Given:

Plate = 16 mm thick

Permissible axial tensile, f_u = 150 N/mm²

1. Bolt Diameter and Area of Bolt at Root:

Let us assume nominal dia 20 mm are used for connections.

Area of bolt at root,

$$A_{nh} = 0.78 \, A_{sb}$$

$$= 0.78 \frac{\pi}{4} d^2$$

$$= 0.78 \frac{\pi}{4} 20^2$$

Answer = 245 mm²

2. Design Strength in Shear:

$$V_{dsb} = \frac{V_{sb}}{\gamma_{mb}}$$

$$V_{nsb} = \frac{f_{ub}}{\sqrt{3}} \left[n_n A_{nb} + n_s A_{sb} \right]$$

$$V_{dsb} = \frac{400}{\sqrt{3}} \left[(1 \times 245) + (0 \times 314) \right] = 56580 \, N$$

$$\gamma_{mb} = 1.25 \text{ as per Table of IS } 800$$

$$f_{ub} = 400 \text{ MPa}$$

Grade 4.6 bolts.

$$n_S = 0$$

$$n_n = 1$$

$$V_{dsb} = \frac{56580}{1.25}$$

$$= 45266 \, N.$$

3. Design Strength in Bearing [V_{dpb}]:

$$V_{dpb} = \frac{V_{npb}}{\gamma_{nub}}$$

$$= \frac{2.5 \, k_b \, dt \, \delta_u}{\gamma_{mb}}$$

Where k_b is the least of the following:

- $\dfrac{e}{3d_o} = \dfrac{1.5[20]}{3\times22} = 0.4545$

- $\dfrac{P}{3d_o} - 0.25 = \dfrac{2.5\times20}{3\times22} - 0.25 = 0.508$

- $\dfrac{f_{ub}}{f_u} = \dfrac{400}{410} = 0.975$

- 1.0.

$$K_b = 0.4545$$
$$V_{npb} = 2.5\times 0.4545 \times 20 \times 16 \times 410$$
$$V_{dpb} = \dfrac{149076}{1.25}$$
$$V_{dpb} = 119260 \text{ N.}$$

Bolt Value is least of above two,

Bolt Value = 45266N

4. Full Strength of Solid Plate Governed by Yielding:

Strength of Solid plate per Pitch Length $T_{dg} = \dfrac{A_g f_y}{\gamma_{m_o}}$

$A_g = P\,t = 50 \times 16$

$A_g = 800$ mm²

$f_y = 250$ MPa.

$f_{Mo} = 1.1$ [as per Table of IS 800-2007]

$T_{dg} = \dfrac{[800]\times250}{1.1}$

= 181818.2 N

No. of bolts required $= \dfrac{\text{strength of solid plate}}{\text{Bolt value}} = \dfrac{181818.2}{45266} \approx 4\,\text{Nos.}$

No. of bolts = 4 Nos.

2. Let us discuss about a single bolted double cover butt joint used to convert boiler plates of thickness 12 mm for maximum efficiency. Use M16 bolts of grade 4.6. Boiler plates are of the 416 grade.

Solution:

Given:

Thickness =12 mm

d = 16 mm

d_o = 18 mm

f_{ub} = 400 N/mm²

f_u = 410 N/mm², t = 12 mm

Since it is double cover butt joint, the bolts are in double shear one section at shank and another at root.

Nominal strength of a bolt in shear,

$$= \frac{400}{\sqrt{3}}\left(\left(1\times\frac{\pi}{4}\times16^2\right)+\left(1\times0.78\times\frac{\pi}{4}\times16^2\right)\right)$$

$$= 82651 \text{ N}$$

Design strength in shear $= \dfrac{82651}{1.25} = 66121 \text{ N}$ ...(1)

Provide cover plates of 8 mm thickness.

Design strength of plate per pitch width,

$$= \frac{0.9\times410(P-18)\times12}{1.25}$$

$$= 3542.4\,[P - 18]$$ ...(2)

Equating (1) and (2) to get maximum efficiency,

3542.5 (P – 18) = 66121

P = 86.67

3. Let us design a lap joint connecting two plates of thickness 10mm and 12mm respectively to carry a factored load of 150 kN. Use 16mm diameter 4.6 grade bolts and Fe410 grade steel.

Solution:

Given:

Thickness= 10mm and 12mm

Load =150 kN

Design a Lab Joint

Connect: Two plates of thickness = 10 mm and 12 mm

Factored Load, P = 150kN

Diameter of bar = 16mm

Grade of bolt = 4.6

Fe410 grade Steel.

Efficiency = ?

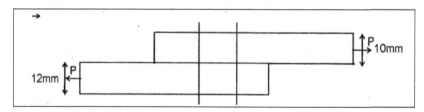

Nominal diameter of the bolt, d = 16 mm

Diameter of the hole, d_0 = 16 + 2 = 18mm

1. Bolt Diameter and Area of Bolt at Root [A_{nb}]

$$A_{nb} = 0.78\left[\frac{\pi[16]^2}{4}\right] = 157\,mm^2$$

2. Strength of 1 Bolt in Single Shear, V_{dsb}

$$V_{dsb} = \frac{V_{nsb}}{\gamma_{mb}}$$

$$V_{nsb} = \frac{\delta_{ub}}{\sqrt{3}}\left[n_n A_{nb}^+ \, n_s A_{sb}\right]$$

As per IS 800 – 2007;

fub = 400 MPa [for M16 and 4.6 grade bolts as per IS 800.2007 –Table].

$$n_n = 1\,;\, A_{nt} = 157\,mm^2,\, n_s = 0$$

$$V_{nsb} = \frac{400}{\sqrt{3}}\left[1\left[157\right]+0\right] = 36259\,N$$

$$V_{dsb} = \frac{36259}{1.25} = 29007\,N$$

3. Design Strength of 1 Bolt in Bearing [V_{dpb}]

$$V_{dpb} = \frac{V_{npb}}{\gamma_{mb}}$$

$$V_{npb} = 2.5\, k_b\, dt\, f_u$$

k_b is smaller of $\dfrac{e}{3\, d_o}, \dfrac{P}{3d_o} - 0.25, \dfrac{S_{ub}}{f_u}$ and 1

Minimum end distance $e = 1.5d = 1.5 \times 16$

$e = 2400 \approx 30$ mm

Minimum Pitch = 2.5 d = 2d [16] $\approx$ 40 mm

$$k_b = \frac{30}{3[18]}, \frac{40}{3[18]} 0.25; \frac{400}{410} = 0.56,\, 0.491,\, 0.976$$

$k_b = 0.491$, d = 16 mm, t = 12 mm, $f_u = 410$ MPa

$V_{npb} = 2.5 \times 0.491 \times 16 \times 10 \times 410$

$V_{npb} = 80524$ N

$$V_{dpb} = \frac{V_{npb}}{\gamma_{mb}} = \frac{80524}{1.25} = 64419.2\,\text{N}$$

Bolt Value = Least of 29007 N and 64419.2 N

Bolt Value = 29007 N

4. Number of Bolts Required

$$N = \frac{150000}{29007} = 5.17 \approx 6\,\text{bolts}$$

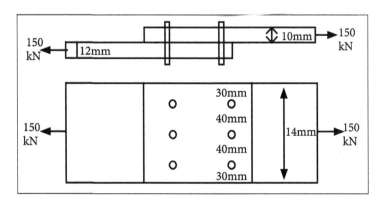

5. Efficiency

$$\eta = \frac{\text{Strength of Joint}}{\text{Strength of solid plate}} \times 100$$

$$= \frac{29007}{\left[\dfrac{f_y A_g}{200}\right]} \times 100 = \frac{29007 \times 6}{\left[\dfrac{250 \times 40 \times 10}{100}\right]} \times 100 = 54.69\%$$

Minimum Pitch = 2.5 × 16 = 40 mm

Provide bolts at P = 40 mm

Check for strength of bolt in bearing:

K_b is the minimum of $\dfrac{e}{3do}, \dfrac{P}{3do} - 0.25\dfrac{f_{ub}}{f_u}, 1.0$

Assuming sufficient 'e' will be provided,

K_{bo} = 0.4907

Design strength of bolt in bearing,

$$\frac{2.5 \times 0.4907 \times 16 \times 12 \times 400}{1.25}$$

$$= 75372\ N > 7.66121\ N$$

Hence the assumption that bearing strength is more than design shear is correct.

Design strength of joint per 40 mm width = 666121 N

Design strength of solid plate per 40 mm width,

$$= \frac{250 \times 40 \times 12}{1.1} = 109091\,N$$

Maximum efficiency of joint $= \dfrac{66121}{109091} \times 100 = 60.61\%$

1.4   Welded Connections: Assumptions and Types

Welded Connection

When two structural members are joined by means of welds the connection is called a welded connection. A few decades ago designers had a feeling that welded connections were less fatigue resistant and that a good quality welded connection could not be made. These negative feelings had a great impact on the use of welding in structures.

But the progress made in welding equipment and electrodes, the advancing art and science of designing for welding and the increasing trust and acceptance of welding have combined to make it a powerful implement for the expanding construction industry.

Structural welding is a process by which the parts that are to be connected are heated and fused, with supplementary molten metal at the joint. A relatively small depth of material will become molten and upon cooling, the structural steel and weld metal will act as one continuous part where they are joined.

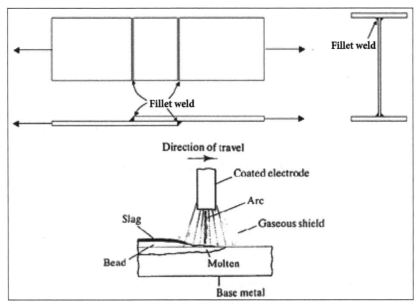

Welded Connection.

Assumptions in the Analysis of Welded Joints

The following assumptions are made in the analysis of welded joints:

- The welds connecting the various parts are homogeneous, isotropic and elastic elements.

- The parts connected by the weld are rigid and their deformations are therefore, neglected.

- Only stresses due to external loads are considered. Effects of residual stresses, stress concentrations and shape of the welds are neglected.

Advantages of Welded Connection

- Welded structures are lighter.

- Welding process is quicker.

- Welding is more adaptable.

- It provides good aesthetic appearance.

Disadvantages of Welded Connection

- Brittle fracture is highly possible.

- It is likely to distort in the process.

- They are over rigid.

- The inspection of welded joints is difficult.

Permissible Stresses in Welds

The permissible stress in welds is based on the American Institute of Steel Construction (AISC) building codes. AISC uses the yield strength to determine the max allowable stress. These are minimum safety factors and maximum stress conditions. These permissible stresses could be used to find the necessary size of the weld. If the fillet weld is loaded in torsion the following equations will then be solved to find either the length or leg height of the weld,

$$\tau = \frac{V}{A} \quad \tau^u = \frac{M \cdot r}{J} \quad J = 707 \cdot h \cdot Ju$$

The primary and secondary shear stress must be combined to obtain a total shear stress this will equal the max allowable shear stress. Then it will be found which is the leg height of the weld. Length of the weld is embedded:

$I = 707 \cdot h \cdot Iu$ For fillet welds in bending,

$$\sigma = \frac{M \cdot y}{I}$$

The stress is again known from the permissible shear stress calculation. The leg length (h) or other weld dimensions can then be solved. Length of weld and/or distance between welds are again embedded in I_u.

Values of I_u can be found in table.

Type of loading	Type of weld	Permissible stresses	Safety factor (n)
Tension	Butt	0.60Sy	1.67
Bearing	Butt	0.90Sy	1.11

Bending	Butt	0.60-0.66Sy	1.52-1.67
Simple compression	Butt	0.06Sy	1.67
shear	Butt or fillet	0.40Sy	1.44

Design of Welded Connections

Fillet welds are most common and used in all structures.

Weld sizes are specified in 1/16 in. Increments.

A fillet weld can be loaded in any direction in shear, compression or tension. However it always fails in shear.

The shear failure of the fillet weld occurs along a plane through the throat of the weld as shown in the figure below.

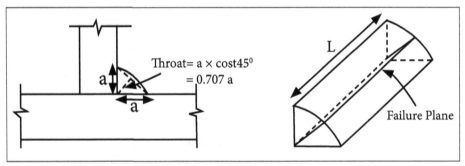

Design of Welded Connections.

Shear stress in fillet weld of length L subjected to load $P = f_v = P/\, 0.707\, a\, L_w$.

If the ultimate shear strength of the weld = f_w.

$$R_n = f_w \times 0.707 \times a \times L_w$$

$$\varphi R_n = 0.75 \times f_w \times 0.707 \times a \times L_w$$

f_w = Shear strength of the weld metal is a function of the electrode used in the SMAW process.

The tensile strength of the weld electrode can be 60, 70, 80, 90, 100, 110 or 120 ksi.

The corresponding electrodes are specified using the nomenclature E60XX, E70XX, E80XX and so on. This is the standard terminology for weld electrodes.

The strength of the electrode should match the strength of the base metal.

If yield stress (σ_y) of the base metal is ≤ 60 - 65 ksi, use E70XX electrode.

If yield stress (σ_y) of the base metal is ≥ 60 - 65 ksi, use E80XX electrode.

E70XX is the most popular electrode used for fillet welds made by the SMAW method.

Table in the AISC Specifications gives the weld design strength,

$$f_w = 0.60 \ F_{EXX}$$

For E70XX, $\varphi \ f_w = 0.75 \ x \ 0.60 \ x \ 70 = 31.5$ ksi

Additionally, the shear strength of the base metal must also be considered:

$$\varphi R_n = 0.9 \ x \ 0.6 \ Fy \ x \ \text{area of base metal subjected to shear}$$

Where,

F_y is the yield strength of the base metal.

For example,

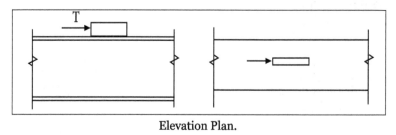

Elevation Plan.

Strength of weld in shear = $0.75 \ x \ 0.707 \ x \ a \ x \ L_w \ x \ f_w$

Strength of base metal = $0.9 \ x \ 0.6 \ x \ F_y \ x \ t \ x \ L_w$

Always check weld metal and base metal strength. Smaller value governs. In most cases the weld metal strength will govern. In weld design problems it is advantageous to work with strength per unit length of the weld or base metal.

Welding

Welding is the method to unite varied items of metal by making a powerful science bond. Bond is achieved by heat or pressure or both. fastening is the most effective and direct means of connecting the metal items. Over several decades totally different fastening techniques are developed to affix metals.

Disadvantages of Welding

- Welding needs bigger talent than bolting and thus needs extremely consummate human resources.

- Improper fastening can distort the members and its alignment and thus needs a lot of concentration.

- The review of weld joints is tougher and cumbersome than fast joints.

- The method of fastening could leave a better residual stress within the material.

- Fastening instrumentality is costlier and needs larger initial investment.

Types of Welds

Butt Welds

A butt weld is made within the cross-section of the abutting plates in a butt or tee joint. Usually the plate edges have to be prepared before welding, shown in figure in some cases if the plate thickness is less than about 5mm, edge preparation may be avoided, shown in figure.

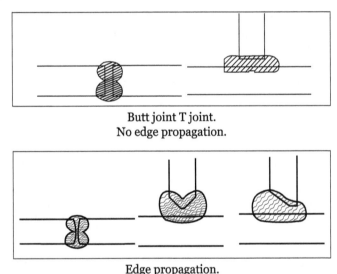

Butt joint T joint.
No edge propagation.

Edge propagation.
Butt weld with full preparation.

The bevelled plate edges in a butt weld may take various geometrical forms, shown in figure.

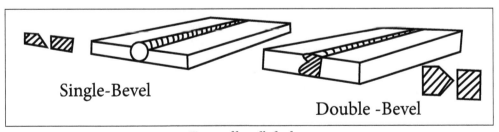

Single-Bevel

Double -Bevel

Types of bevelled edges

For Butt Welds a Distinction is made Between

Full penetration butt weld in which there is a complete penetration and fusion of weld and parent metal throughout the thickness of the joint, shown in figure.

Partial penetration butt weld, in which there is a weld penetration less than the full thickness of the joint, shown in figure.

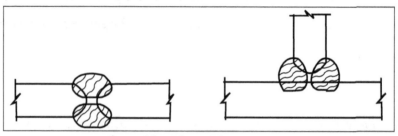

Butt welds with partial penetration.

Fillet Welds

A fillet weld is a weld of approximately triangular cross-section applied to the surface profile of the plates. No edge preparation is required. Therefore fillet welds are usually cheaper than butt welds.

According to the relative position of the parts to be welded, there are three types of fillet weld applications:

Lap joint, in which the parts welded are in parallel planes, shown in figure (a).

Tee or cruciform joint, in which the parts welded are more or less perpendicular to one another, shown in figure (a).

Corner joint in which the parts are also more or less perpendicular to one another, shown in figure (c).

To improve the strength and stiffness of the joint, the outer corner is normally butt welded, shown in figure.

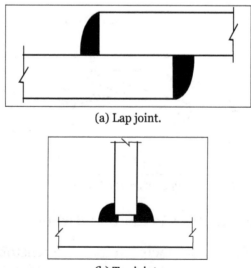

(a) Lap joint.

(b) Tee joint.

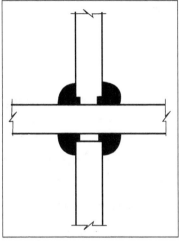

Cruciform joint.

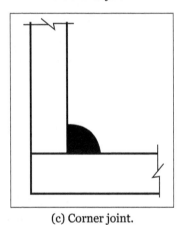

(c) Corner joint.

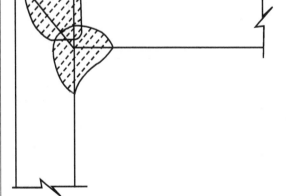

Corner joint with butt and fillet weld.

Fillet welds that can be laid in a single run are particularly economic. In the workshop 8mm welds are often possible but if site welding is to be used this figure might be reduced 6mm.

1.4.1 Design of Fillet Welds

Few terms are used while designing a fillet weld as follows:

1. Size of Fillet Weld

The length of the sides of the largest right angled triangle in the cross-section of the fillet weld is denoted as the size of fillet weld as shown in table (1). Usually the perpendicular sides of such a right angled triangle are equal and the size of a fillet weld may be specified by one dimension only. The size of the fillet weld should not be less than the minimum values given in table (2) as per I.S. 816-1969.

Table (1) Supplementary drawing representation:

Form of Weld	Section	Drawing Representation
Flush butt		Flush square butt.
Machine finish		Double V - butt finish flush on other side.
Grinding finish		Double bevel butt finished flush on arrow side. Arrow mark on one section only indicates the section prepared for joint.
Convex fillet		
Concave fillet using field weld		Black circle indicates fillet weld.
Weld all round		Circle indicates fillet weld all-round the section.

Table (2) Minimum Size of Fillet Weld:

Thickness of thicker part	Minimum size
Up to and including 10 mm	3 mm
Over 10 mm up to and including 20 mm	5 mm
Over 20 mm up to and including 32 mm	6 mm
Over 32 mm up to and including 50 mm	8 mm first run 10 mm minimum

The maximum size of the fillet weld is also specified. The maximum size of the fillet weld applied to the square edge of a plate or shape should be 1.5 mm less than the nominal thickness of the edge. The size of the fillet weld used along the toe of an angle or the rounded edge of a flange should not exceed three fourths the nominal thickness of an angle or flange leg.

2. Throat of Fillet Weld

The throat of a fillet is the length of perpendicular from the right angle corner to the hypotenuse as shown in table (1). The effective thickness of throat is calculated as,

Throat thickness =Kx fillet size

The value of K depends upon the angle between the fusion faces as given in table (3).

Table (3) Values of K:

Angle between fusion faces	60-90°	91-100°	101-106°	107-113°	114-120°
Constant K	0.7	0.65	0.6	0.55	0.5

In most cases, a right-angled fillet is used, for which K = $1/\sqrt{2}$ or 0.7.

3. Effective Length of Fillet Weld

The effective length of a fillet weld is equal to its overall length minus twice the weld size. The effective length of a fillet weld designed to transmit loading should not be less than four times weld size. Only the effective length is shown on the drawing and the additional length (i.e. 2 x weld size) is provided by the welder.

4. End Return

The fillet weld terminating at the end or side of a member should be returned around the corner whenever practicable for a distance not less than twice the weld size as shown in figure (a).

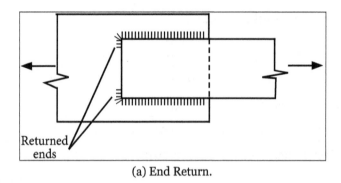

(a) End Return.

This provision applies in particular to fillet welds in tension connecting beam seatings, brackets, etc.

5. Overlap

The overlap in a lap joint should not be less than five times the thickness of the thinner plate as shown in figure (b).

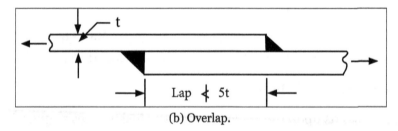

(b) Overlap.

6. Side Fillet

In a lap joint made by a side or longitudinal fillet weld, the length of each fillet weld should not be less than perpendicular distance between them. The perpendicular distance between the side fillets should not exceed sixteen times the thickness of the thinner part connected. Therefore, shown in figure(c), l is not lesser than b and b is not greater than 16t, where t = thickness of thinner plate. If b exceeds this limit then additional end fillet, plug or slot weld is provided to prevent buckling or separation of the parts.

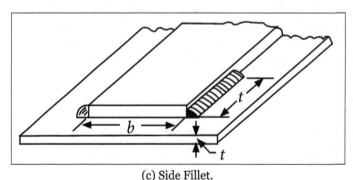

(c) Side Fillet.

7. Intermittent Fillet Weld

Intermittent fillet weld may be used when the length of the smallest size fillet weld required to transmit stress is less than the continuous length of the joint. Any section of an intermittent fillet weld should have an effective length of not less than four times the weld size or 40 mm, whichever is greater.

The clear spacing between the ends of the effective lengths of intermittent fillet weld carrying stresses should not exceed 12 t for compression and 16 t for tension and in no case should be more than 20 cm; where t= thickness of thinner part joined.

A chain intermittent welding is to be preferred to staggered intermittent welding. When staggered intermittent welding is used, the ends of the component part should be welded on both ends.

8. Single Fillet Weld

A single line of fillet weld, when used, should not be subjected to bending about its longitudinal axis.

9. Permissible Stress and Strength of Filet Weld

The fillet welds are designed for the shear stress at the minimum section, i.e. the throat of the weld. The permissible stress in the fillet weld is 1100 kgf/cm² or 108 MPa as per IS: 816-1969. The shear strength of a fillet weld is given by the following equation:

$$P = p_q \times l \times t$$

Where,

P = Strength of the joint

p_q = Permissible stress

l = Effective length

t = Throat thickness = $K \times s$

s = Weld size

K = Constant as per table (2)

For the most common case, i.e. welded surfaces meeting at $60° - 90°$,

$$t = 0.7 \times s$$

or,

$$P = 0.7 \times p_q \times l \times s$$

The permissible stresses in shear and tension are reduced to 80% for field welds made during erection. The permissible stresses are increased by 25% if the wind or earthquake load are taken into account. However, the size of the weld should not be less than the size required when the wind or earthquake load is considered or neglected.

Design of Butt Weld

The strength of a butt weld is taken equal to the strength of parts joined if full penetration of the weld metal is ensured. A complete penetration of the weld metal can be ensured in the case of double-V, double-U, double-.1 and double-bevel joint.

In case of single-V, -U. -J and bevel joints, the penetration of the weld metal is generally incomplete and the effective throat thickness is taken as (5/8) x thickness of thinner part connected. The change in thickness while joining unequally thick plates should be gradual. A tapper not exceeding 1 in 5 is provided, when the difference in thickness of the parts exceeds 25% of the thickness of the thinner part or 3 mm. whichever is greater.

The permissible stress as per IS: 816-1969 for butt weld is taken to be the same as that of the parent metal.

1.4.2 Intermittent Fillet Weld

Design of Intermittent Fillet Welds

Intermittent fillet welds are provided to transfer calculated stress across a joint when the strength required is less than that developed by a continuous fillet weld of the smallest practical size, Example. in case of connections of stiffeners to the web of plate girders. The fillet weld length required is computed as a continuous fillet weld and a chain of intermittent fillet welds of total length equal to the computed length, with spacing as per I.S.

A specification is provided, as shown in figure. Intermittent fillet welds as shown in figure (a) are structurally better than those shown in figure(b).

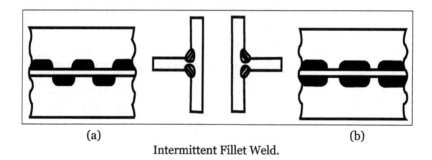

(a)   (b)

Intermittent Fillet Weld.

The question of whether and to what extent intermittent welds should be used involves consideration of the following points:

- Intermittent fillet welds are not economical unless the weld is of minimum size. A smaller fillet weld of a longer length is usually more economical for the same strength. This is because the strength of a fillet weld increases directly with size but the weight of the weld metal increases with the square of the size.

- If automatic welding is to be used the weld should be continuous.

- If the structure is exposed, the use of continuous welds may be preferable as they are conducive to greater ease of maintenance and longer life of structure.

- If severe dynamic loads act on the structure, intermittent welds must not be used.

Design

- The size of weld is assumed and the total effective length of the intermittent fillet weld required is computed.

- 2. Any intermittent fillet weld section should have a minimum effective length of four times the size of the weld with a minimum of 40 mm, except for plate girders.

- The clear spacing between an intermittent fillet weld should not exceed 12t for compression and 16t for tension and should in no case be more than 200 mm.

- At the ends, the longitudinal intermittent fillet weld should be of a length not less than the width of the member or else transverse welds should also be provided. If transverse welds are also provided along with longitudinal intermittent fillet welds, the total weld length at the ends should not be less than twice the width of the member.

1.4.3 Plug and Slot Weld

Plug and slot welds are used in addition to fillet welds when sufficient welding length is not available along the edges of the members. They are also provided in some cases to prevent the buckling of weld plates.

A slot is cut in one of the overlapping members and the welding metal is filled in the slot. If the slot is small and completely filled with weld metal, it is known as a plug weld, but if only the periphery of the slot is fillet welded then it is known as a slot weld.

The following specifications are for the design of plug or slot weld as per IS: 1816-1969:

- The width or diameter of the slot should not be less than three times the thickness of the part in which the slot is formed or 25 mm, whichever is greater.

- Corners at the enclosed ends should be rounded to a radius not less than one and a half times thickness of the upper plate or 12 mm, whichever is greater.

- The distance between the edges of the plates and the slot or between the edges of adjacent slots should not be less than twice the thickness of the upper plate.

- Plug and slot welds are designed for shear stress acting at throat.

Strength of plug or slot weld = Permissible stress x cross-sectional area at throat.

The permissible stress is taken as 1100 kgf/cm^2 or 10^8 MPa.

- Plug welds are not designed to carry loads.

1.4.4 Failure of welded joints

Butt Welds

When the butt weld is reinforced on both the sides of the plate, the section through the weld is increased to such an extent that it is unlikely for failure to occur in the weld and the fracture normally occurs some distance away from figure below.

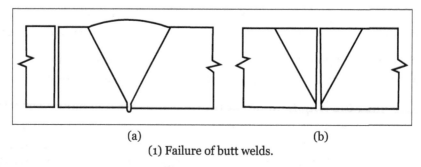

(a)   (b)

(1) Failure of butt welds.

The reinforcement acts as a supporting rib which inhibits necking in the immediate vicinity of the weld.

If the weld is ground flush with the surface of the plate, the position of the fracture depends upon the relative strength of the plate and weld metal. If the tensile strength or the yield point is of Butt Welds lower for the weld metal than for the plate, failure takes place through the centre of the weld shown in figure (b). But if the tensile strength and yield point of the weld metal are higher, failure takes place in the plate away from the weld.

Failure in the weld junction is quite unusual. Under conditions of free bending across a reinforced butt weld, the stiffening effect of the reinforcement inhibits failure at the joint. But since stress concentration occurs at the weld junction, due to change of the section, there is some tendency for a crack to start at the weld junction itself with a flush butt weld, failure under such a condition occurs in the middle of the weld. If there is a wide difference in the tensile strength or yield point between the two metals, the failure may occur at the junction.

End Fillet Weld

The plane of the fracture in a normal profile convex end fillet weld shown in the figure (a), is along the diagonal from the root of the fillet figure (b). When the fillet is subjected to shear in addition to the tensile stress, the position of the line of fracture departs from the diagonal according to the relative magnitude of the two stresses as shown in the figure (c). If the fillet legs are unequal, a fracture usually occurs near the shorter leg a shown in the figure (d).

If the tensile strength of the weld metal is considerably greater than that of the plate, the fillet may remain intact and be pulled right out of the plate as shown in the figure (e). With all end fillet welds failures occur abruptly, after a small amount of deformation.

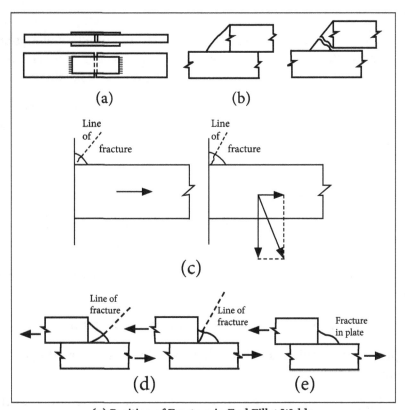

(2) Position of Fracture in End Fillet Welds.

Side Fillet Weld

In a side convex fillet weld subjected to shear stress along the weld, failure occurs in the throat of the weld. The break commences at the toe of the fillet at one or both the ends of the weld and as it progresses, the plane of fracture rotates in figure. The failure is gradual and considerable deformation of the fillet and of the plates takes place before the final fracture.

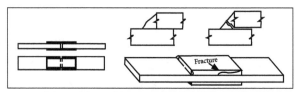

(3) Position of fracture in side fillet welds.

1.4.5 Welded Joints vs Bolted and Riveted Joints

- Welded joints are economical. This is because splice plates and bolt/rivet materials are eliminated. Also, the gusset plates required are of a smaller size because of the reduced connection length. Labour cost is also less as only one person is required to do the welding whereas at least two persons are required for bolting and four for riveting.

- Welded structures are more rigid as compared to bolted/riveted joints. In bolted/riveted joints, cover plates, connecting angles, etc., deflect along with the member during load transfer and make the joint more flexible.

- Due to the fact that the strength of a welded joint is the same as that of the parent metals, even a smallest piece of the metal which otherwise is a scrap can be used, bringing overall economy.

- With welding it has become possible to connect tubular sections, which are structurally very economical.

- Due to the fusion of two metal pieces jointed, a continuous structure is obtained, which gives a better architectural appearance than bolted/riveted joints.

- Alterations can be done with less expense in case of welding as compared to bolting/riveting.

- The process of welding is quicker in comparison to bolting/riveting.

- The process of welding is silent, whereas in the case of riveting a lot of noise is produced.

- In welding less safety precautions are required for the public in the vicinity, whereas a hot rivet may toss and injure the persons working.

- As splice plates, bolts/rivets, etc., are not used, the details and drawings of welded structures are easier and less time consuming.

- The efficiency of welded joint is more than that of a bolted/riveted joint. In fact, a proper welded joint may have 100% efficiency.

- Members to be jointed may distort due to the heat during the welding process, whereas there is no such possibility in bolted/riveted joints.

- The possibility of a brittle fracture is more in the case of welded joints as compared to bolted/riveted joints.

- The inspection of welded joints is difficult and expensive, whereas bolted/ riveted joints can be inspected simply by tapping the joint with a hammer.

- A more skilled person is required to make a welded joint as compared to a bolted/ riveted joint.

Problems

1. A tie member of a roof truss consists of 2 ISA 10075, 8mm. The angles are connected to either side of a 10 mm gusset plates and the member is subjected to a working pull of 300 kN. Let us design the welded connection and assume connections are made in the workshop.

Solution:

Given:

Working Load = 300 kN

∴ Factored Load = 300 × 1.5 = 450 kN

Thickness of Weld

- At the rounded toe of the angle section, size of weld should not exceed = $\frac{3}{4}$ × thickness.

- At top (Refer figure) the thickness should not exceed,

    s = t – 1.5 = 8 – 1.5 = 6.5 mm

Hence provide s = 6 mm, weld.

Each angle carries a factored pull of 450/2= 225 kN

Let L_w be the total length of the weld required.

Assuming normal weld, t = 0.7 × 6 mm

∴ Design strength of the weld = $L_w t \dfrac{f_u}{\sqrt{3}} \times \dfrac{1}{1.25}$

$$= L_w \times 0.7 \times 6 \times \frac{410}{\sqrt{3}} \times \frac{1}{1.25}$$

Equating it to the factored load, we get

$$L_w \times 0.7 \times 6 \times \frac{410}{\sqrt{3}} \times \frac{1}{1.25} = 225 \times 10^3$$

∴ L_w – 283 mm

Centre of gravity of the section is at a distance 31 mm from top.

Let L_1 be the length of top weld and L_2 be the length of lower weld. To make centre of gravity of weld to coincide with that of angle,

$$L_1 \times 31 = L_2(100-31)$$

$$\therefore L_1 = \frac{69}{31}L_2$$

$$L_1 + L_2 = 283$$

i.e.,

$$L_2\left(\frac{69}{31}+1\right) = 283$$

or,

$$L_2 = 87 \text{ mm}$$

$$\therefore L_1 = 195 \text{ mm}$$

Provide 6 mm weld of L_1 = 195 mm and L_2 = 87 mm as shown in the figure.

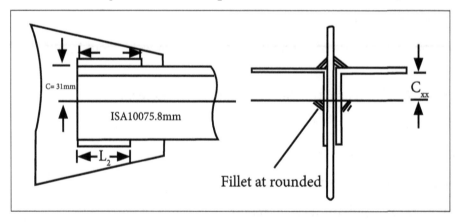

2. Let us design a ISA90 × 90 × 10mm connected to a gusset plate 12mm thick by welding. The member carries an axial factored load of 200 kN. Use Fe410 grad steel.

Solution:

Given:

Design Weld Connection

ISA 90 × 90 × 10mm.

Connect to guest plate = 12 mm thickness by Welding.

Factored Load = 200 kN

Weld Connection an all three sides

Use Fe410 grade steel.

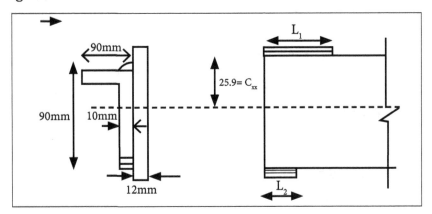

1. Design Load

Factored Load = 200 kN

2. Size and Throat Thickness of Weld

(i) Maximum size of weld at rounded toe of angle,

> = 3/4 [TK of angle]

> = 3/4 [10] = 7.5mm

(ii) Maximum size of weld = TK of thinner Plate,

> = 1.5m

> = 10 – 1.5 = 8.5mm

(iii) Minimum size = 5 mm [as per Table 2 of IS 800 for H of thicker part between 10 to 20 mm.]

Let us choose the weld size as 7.5 mm

Throat thickness (Tk),

> = 0.7[Size of Weld]

> = 0.7 × 7.5 = 5.25mm

3. Strength of Weld

Pull transmitted by 1mm length of weld = $A_g f_{wd}$ [as Per Cl. 10.5.7.1.1 of IS800:2007]

$$f_{wd} = \frac{f_{wn}}{\gamma_{mw}} = \frac{(fu/\sqrt{3})}{\gamma mw}$$

$$= \left[\frac{410/\sqrt{3}}{1.25}\right] \Rightarrow 189.38\,\text{MPa}$$

Strength/mm of weld $= \dfrac{1[0.7\times7.5]\left[410\sqrt{3}\right]}{1.25} = 994\,\text{N} = 994\text{N}$

4. Length on Weld

L_w = Total Length of weld on top and bottom on each angle.

Force transmitted by each Angle = 200 kN

Factored Load = Design Strength of Weld

$100 \times 3 = 994\,L_w$

$L_w = 201.6 \approx 201\,\text{mm}$

5. Length on Each Side of One Angle

Centre of Gravity of ISA 90 × 90 × 10mm lies at 25.9 mm from top.

Let L_1 be the Length of weld at top and L_2 be the Length of weld at bottom.

Forces above C.G of angle = Forces below C.G of angle.

$$L_1[25.9] = L_2[90 - 25.9]$$
$$L_1 = \frac{90-25.9}{25.9}L_2$$
$$L_1 = 2.47\,L_2$$
$$L_\omega = L_1 + L_2$$
$$201 = 2.47\,L_2 + L_2$$
$$201 = 3.47\,L_2$$
$$L_2 = 201/3.47 = 57.9\,\text{mm}$$
$$L_1 = 201 - 57.9$$
$$L_1 = 143.1\,\text{mm}$$

∴ Provide 7.5 mm weld of Length,

L_1 = 143.1 mm at top and Length,

L_2 = 57.9 mm at bottom on each of the angle.

3. Let us design a lap joint of 100 mm × 10 mm plate to be welded to another plate 150 mm × 10 mm by the fillet welding on three sides. The size of the weld is 6 mm. Take allowable tensile stress in plate equal to 150 N/mm² and allowable stress in weld as 110 N/mm².

Solution:

Given:

Lap joint, A_g =100 mm × 10 mm

F_y =250

1. Design Load

Strength of Plate $\dfrac{A_g f_y}{\gamma m_o} = \dfrac{[100 \times 10] \times 250}{1.1}$

As for Table of IS 800-2007.

Strength of Plate = 227272.7 N

2. Size and Threat Thickness of Weld

Size of Weld = 6 mm

Throat Thickness = 0.7 × Size of Weld,

= 0.7 × 6 = 4.2mm

3. Strength of Weld Per 1 mm

Strength of Weld/mm Length = A_g [f_{wd}]

$$f_{wd} = \frac{(fu / \sqrt{3})}{\gamma m_w}$$

$$= \frac{410}{1.25 \times \sqrt{3}} = 189.2820 \text{ N/mm}^2$$

Strength of Weld /mm = 1 × 4.2 × 189.282 = 795.98 N.

4. Length of Weld Required

L_w × 794.98 = 227272.7

L_w = 286 mm

But length for fillet weld available on sides for 100 mm plate,

$$= 2 \times [100 - 2 \times 6] = 176 \text{ mm}$$

∴ Force transmitted by 176 mm length of fillet,

$$\text{Weld} = 176 \times 285 = 50336 \text{ N}$$

The remaining strength is to be transmitted through slot welds.

∴ Strength to be transmitted through Slot Welds = 227272.7

Area of Slot Weld Requireds

$$\text{Area of Slot Weld} \times \frac{250\sqrt{3}}{\gamma m_w} = 176936.7$$

Area of Slot Weld = 1532mm²

Hence let us provide two slot welds of size 50 × 20 mm each giving an area.

$$= 2[50 \times 20] = 2000$$

4. Let us design a tie member consisting of angle section 80 mm × 50 mm × 8 mm welded to a 8 mm gusset plate.

Solution:

Given:

Angle section = 80 mm × 50 mm × 8 mm

Welded= 8 mm

Step 1: Determination of full strength of the member ISA 80 mm × 50 mm × 8 mm. From steel table c/s area of ISA 80 × 50 × 8 mm = 978 mm².

Assume yield strength of 250 MPa for the member partial safety factor for angle section as per Table IS 800-2007 is given by,

$$\gamma_{mo} = 1.1$$

∴ Design Strength of the member

$$= \frac{A_g fy}{\gamma_{mo}}$$

$$= \frac{978(250)}{1.1}$$

$$= 222273 \text{ N}$$

Step 2: Size and throat thickness of Weld.

(i) Maximum size of weld at rounded toe of angle,

$$= \frac{3}{4} \text{ (Thickness of Angle)}$$

$$= \frac{3}{4} = 6 \text{ mm}$$

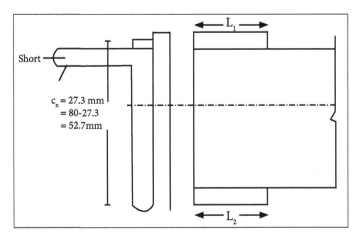

(ii) Maximum Size of Weld = Thickness of thinner plate – 1.5 mm

$$= 8 - 1.5 = 6.5 \text{ mm}$$

(iii) Minimum size of weld = 3 mm

For thickness of thicker up to and including 10 mm.

∴ let the Weld size as 6 mm

Throat thickness = 0.7 (size of weld) = 0.7 (6) = 4.2 mm

Step 3: Strength of weld per 1 mm length

$$f_{wd} = \frac{f_{wn}}{\gamma_{mw}} = \frac{fu/\sqrt{3}}{\gamma_{mw}}$$

Assume fu = 250 MPa,

$$\gamma_{mw} = 1.25$$

$$f_{wd} = \frac{250/\sqrt{3}}{1.25} = 115.47 \text{ Mpa}$$

∴ Strength/mm of Weld = 1(0.7) (6) (115.47) = 484.99 ≈ 485 N

Step 4: Length of Weld required.

Equate the design strength to factor load.

We get,

$$L_w (485) = 222273 \text{ N}$$
$$L_w = 458.31 \text{mm} \approx 459 \text{ mm}$$

Step 5: Length of each side of one angle.

$$L_1 (27.3) = L_2 (80 - 27.3)$$
$$L_1 = 1.9342 \ L_2$$

Also,

$$L_1 + L_2 = L_w = 459 \text{ mm}$$

$$1.93 \ L_2 + L_2 = L_2 \ (1.93 + 1) = 459$$

$$L_2 = \frac{459}{2.93} = 237.77 \approx 238 \text{ mm}$$

$$L_1 = 1.93(238) = 459.34 \approx 460 \text{ mm}$$

∴ Provide 6 mm weld of length L_1 = 460 mm at top and length

$$L_2 = 238 \text{ mm at bottom.}$$

5. Let us design an I.S.L.C. 300 at 331 N/m used to transmit a force of 600 kN. The channel section connected to a gusset plate 10 mm thick overlap is limited to 350 mm. Also use a slot weld if required.

Solution:

Given:

  Force = 600 kN

  Thickness = 10 mm

Relevant properties of I.S.L.C. 300 at 331 N/m are,

  A = 4211 mm²

  T = 11.6 mm

  t = 6.7 mm

As slot welds are required, the maximum size of weld,

= 6.7 – 1, 5

= 5.2 mm

Provide 5 mm-size weld.

Throat thickness = 0.7 × 5 = 3.5 mm

Strength of weld/mm length = 3.5 × 108 = 378N

Length of weld required = $\dfrac{600 \times 1000}{378}$ = 1587.30mm

The maximum length of the weld that can be provided is,

300 + 2 × 350 = 1000 mm (1590 mm).

Hence, the length of the weld will be provided with slot welds

Provide Slots of width 25 mm (3t = 3 × 6.7 = 20.1 mm or 25 mm whichever is greater).

Let length of the slot be x,

1590 = 2 × 350 + 300 + 4x

x = 147.5 mm = 150 mm

Provide 150mm long fillet welds in slots as shown in figure,

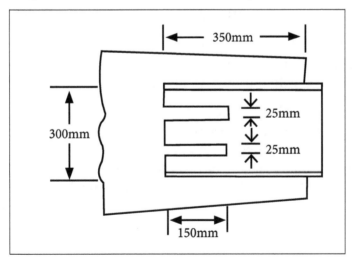

Tension Members and Compression Members

2.1   Tension Members and Types

The Tension member could be through of a linear member that carries axial pull. The members endure extension as a result of this axial pull. This can be the common sorts of force transmitted in the structural system. Tension members are economical since the complete cross section carries uniform stress in contrast to flexural members.

The strain members don't buckle even once stressed on the far side the elastic limit. Therefore, the look isn't accomplished by the sort of section used i.e., Plastic, Compact or Semi-compact. A number of the common samples of tension members in structures are Bottom chord of pin articulate roof trusses, bridges, line and communication towers, wind bracing system in multistory buildings, etc.

Behaviour of Tension Members

The load deformation behaviour of members subjected to uniform tensile stress is comparable to the load-deflection behavior of the corresponding basic material. The higher yield purpose is unified with the lower yield purpose for convenience.

The material shows a linear elastic behavior within the initial region (O to A). The fabric undergoes enough yielding in portion A to B. Additional deformation ends up in a rise in resistance wherever the fabric strain hardens (from B to C). The fabric reaches its final stress at portion C. The strain decreases with increase in additional deformation and breaks at D. The high strength steel members don't exhibit the well outlined yield purpose and therefore the yield region. For such materials the 0.2 % proof stress is typically taken because the yield stress (E).

Types of Tension Members

The types of structure and method of end connections determine the type of a tension member in structural steel construction. Tension members used may be broadly grouped into four groups:

- Wires and cables.

- Rods and bars.

- Single structural shapes and plates.

- Built-up members.

1. Wires and Cables

The wire types are used for hoists, derricks, rigging slings, guy wires and hangers for suspension bridges.

2. Rods and Bars

The square and round bars are quite often used for small tension members. The round bars with threaded ends are used with pin-connections at the ends instead of threads.

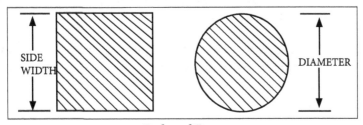

Rods and Bars.

The ends of rectangular bars or plates are enlarged by forging and bored to form eye bars. The eye bars are used with pin connections. A rods and bars have the disadvantage of inadequate stiffness resulting in noticeable sag under the self-weight.

3. Single Structural Shapes and Plates

The single structural shapes, i.e. angle sections and tee-sections as shown in figures below, are used as tension members. The angle sections are considerably more rigid than the wire ropes, bars and rods. When the length of tension member is too long, then the single angle section also becomes flexible.

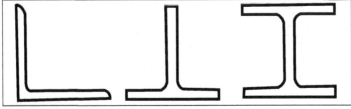

Single Structural Shapes and Plates.

The single angle sections have the disadvantage of eccentricity in each plane in a riveted connection. The channel section has eccentricity in one axis only. Single channel sections have high rigidity in the direction of web and low rigidity in the direction of flange. A occasionally, I-sections are sued as tension members. The I-sections have more rigidity and single I-sections are more economical than built up sections.

4. Built-up Sections

Two or more members are used to form built up members. When the single rolled steel section cannot furnish the required area, then built-up sections are used. The double angle sections of unequal legs shown in the figure are extensively used as tension members in the roof trusses. The angle sections are placed back to back on two sides of a gusset plate.

When both the angle sections are attached on the same side of the gusset, then built-up section has eccentricity in one plane and is subjected to tension and bending simultaneously. The two angle sections may be arranged in the star shape. The star shape angle sections may be connected by batten plates.

The batten plates are alternatively placed in two perpendicular directions. The star arrangement provides a symmetrical and concentric connection. Two angle sections as shown in the figure (a) are used in the two-plane trusses where two parallel gussets are used at each connection.

Two angle sections as shown in figure (b) have the advantage that the distance between them could be adjusted to suit connecting members at their ends. Four angle sections as shown in figure (c) are also used in the two-plane trusses. The angles are connected to two parallel gussets. For angle, sections connected by plates as shown in figure (d) are used as tension members in bridge girders.

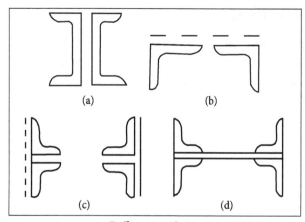

(a)   (b)

(c)   (d)

Built-up sections.

A built-up section may be made of two channels placed back to back with a gusset in between them. Such sections are used for medium loads in a single plane-truss. In two-plane trusses, two channels are arranged at a distance with their flange turned inward.

It simplifies the transverse connections and also minimizes lacing. The flanges of two channels are kept outwards, as in the case of chord members or long span girders, in order to have greater lateral rigidity. The heavy built-up tension members in the bridge girder trusses are made of angles and plates. Such members can resist compression and reversal of stress takes place.

Type of Steel Section Used as Tension Member

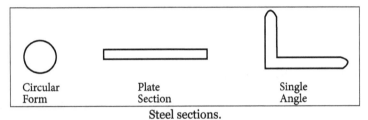

Circular Form	Plate Section	Single Angle

Steel sections.

The tension members may be made of single structural shapes. The standard structural shapes of typical tension members are:

- Angle section.

- Tee section.

- Channel section.

- Box section.

- I section.

- Tubular section.

The sections may be built up using a number of top structural shapes. Single angle members are economical however the connection produces eccentric force within the member. These are usually used in towers and in trusses. Double angle members are more rigid than single angle members. They are used in roof trusses.

Since there exists a gap of about 6 to 10 mm between the two members, they are generally interconnected at regular intervals so that they act as one integral member. Within the members of bridge trusses the tensile forces developed are very large and hence require more rigid members. In these structures single channel, single I-section, built-up channels or built-up I-sections are usually used.

Tension member is the lowest of the following:

- Design strength due to yielding of gross section T_{dg}.

- Rupture strength of critical section, T_{dn}.

- The block shear tab.

Types of tension steel members:

- Carbon Steel.

- High Strength Carbon Steel.

- Medium and High Strength Micro Alloyed Steel.

- High Strength Quenched and Tampered Steels.

- Weathering Steels.

2.1.1 Net Cross-Sectional Area

A tension member is designed for its net sectional area at the joint. When a tension member is spliced or joined to a gusset plate by rivets or bolts, the gross sectional area is reduced by rivet holes.

The net sectional area is calculated for various cases as per IS: 800-1984 as follows:

(i) For plates In case of chain riveting, the net sectional area at any section is given by,

$$A_{net} = b \times t - n \times d \times t$$

or,

$$A_{net} = t \times (b - n \times d) \qquad \qquad ...(1)$$

Where,

A_{net} = net cross-sectional area along rivet chain.

b = Width of plate.

t = Thickness of plate.

d = Gross diameter of rivet holes.

n = No. of rivets at the section.

In case of a zig-zag or diagonal chain of holes, the net cross-sectional area along a chain of rivets is increased by an amount equal to,

$$\frac{s^2 \times t}{4_g}$$

Where,

s - Staggered pitch, i.e. the distance between any two consecutive rivets measured parallel to the direction of the stress in the member.

g - Gauge distance, i.e. the distance between the two consecutive rivets in a chain measured at right angles to the direction of stress in the member.

Thus, in figure (a) net area along section ABCDE,

$$A_{net} = t \times \left[b - n \times d + \left(\frac{s_1^2}{4 \times g_1} + \frac{s_2^2}{4 \times g_2} \right) \right]$$ ...(2)

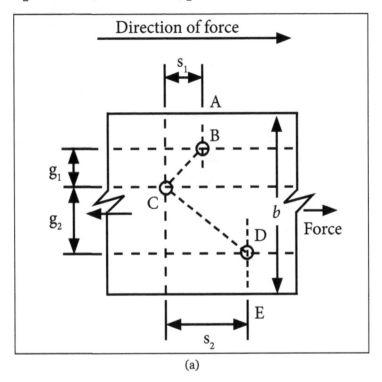

(a)

Where,

n = Number of rivet holes at the section = 3 at section ABCDE

s_1, s_2 = Staggered pitch.

g_1, g_2 = Gauge distance.

For section x-x in figure (b), n = 5,

$$A_{net} = t \times \left(b - 5 \times d + 4 \times \frac{s^2}{4 \times g} \right)$$

For section y-y in figure (b), n = 4,

$$A_{net} = t \times \left(b - 4 \times d + 2 \times \frac{s^2}{4 \times g} \right)$$

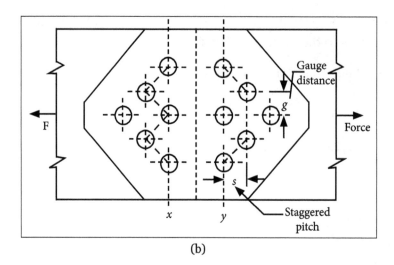

(b)

All such sections are checked for getting the critical or minimum cross sectional area of the plate. The above procedure for the calculation of net area is also applied for angle - sections joined on both legs. The gauge distance in this case is measured along the Centre of the thickness of the angle.

(ii) For single Angle Connected by One Leg Only (figure c)

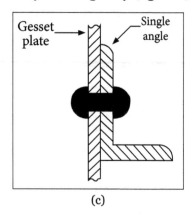

(c)

Net effective area is given by,

$$A_{net} = A_1 + A_2 + k_1 \qquad\qquad ...(3)$$

Where,

A₁ = Net cross-sectional area of the connected leg.

A_1 = Net cross-sectional area of the connected leg.

A_2 = Gross cross-sectional area of unconnected leg.

And,

$$k_1 = \frac{3 \times A_1}{3 \times A_1 + A_2}$$

The area of a leg of an angle,

= Thickness of angle x (length of leg - ½ x thickness of leg).

(iii) For pair of angles placed back-to-back (or a single tee) connected by only one leg of each angle (or by the flange of a tee) to the same side of a gusset plate (figure d).

The net effective sectional area in this case is given by,

$$A_{net} = A_1 + A_2 + k_2 \qquad \qquad ...(4)$$

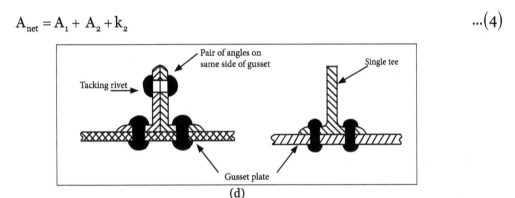

(d)

Where,

A₁ = Net sectional area of the connected legs (or flange of the tee).

A₂ = Area of outstanding legs (or web of the tee).

And,

$$k_2 = \frac{5 \times A_1}{5 \times A_1 + A_2}$$

The area of a web of a tee = Thickness of web x (Depth — Thickness of flange).

The outstanding legs of the pair of angles should be tacked by rivets at a pitch not exceeding 1 m.

(iv) For double-angles or tees carrying direct tension placed back to back and connected to each side of a gusset or to each side of a rolled section (figure e).

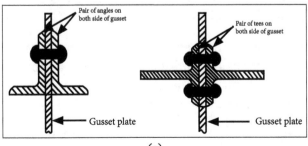

(e)

The net effective sectional area in this case is given by, A_{net} = Gross area — Deduction for holes provided that the angles or the tees are tack-riveted along their length at a pitch not exceeding 1 m. When two angles or tees placed back-to-back are not tack-riveted the provisions under (iii) and (iv) above do not apply and each angle or tee is designed as a single angle or tee connected to one side of the gusset.

Types of Failure

A tension member may fail in any of the following modes and the typical design problem is to select a member with sufficient cross-sectional area to take care of them.

Gross Section Yielding

Considerable deformation of the member in longitudinal direction may take place before it fractures, making the structure unserviceable.

Net Section Ruptures

The rupture of the member when the net cross section of the member reaches the ultimate stress.

Block Shear Failure

A segment of block of material at end of member shears out due to the possible use of high bearing strength of the steel and high-strength bolts resulting in smaller connection length. The factored design tensile load T in the member should be less than the design strength T_d of the member. The design strength of the member under the axial tensile load is the lowest of the design strength due to yielding of gross section T_{dg}, rupture of critical (net) section T_{dn} and block shear T_{db}.

There are two types of failure mechanisms governing the strength requirements of tension members. The first is yielding on the gross area and the second is fracture on the net section. In case of failure by yielding of gross area of the member, it may become unserviceable due to its excessive elongation.

Fracture on the net section on other hand, refers to tearing or fracture of the section perpendicular from the direction of force through the reduced cross-sectional area of a member, typically across the bolt holes.

As the yielding of gross area of the member occurs before the rupture load is reached, the yield load generally governs the limiting lo ad. There are three other modes of failure or limit states for bolted connections in tension. Since the connection is in tension, the bolts can shear, thus can be a shear failure of the bolts.

Also since the bolts bear on the tension member so there can be a bearing failure of the bolts on the tension member. There can also be a block shear failure where a block of the tension member tears away. Some portions of this block are in tension and some are in shear so this is a combination of shear and tension that result in this type of failure.

2.1.2 Slenderness Ratio

The slenderness ratio is defined as the ratio of effective length or height of a structural member to its least radius of gyration,

$$\text{Slenderness ratio} = \frac{\text{Effective height}}{\text{Least radius of gyration}}$$

Apart from strength requirement, the tension members have to be checked for low stiffness by stipulating the limiting maximum slenderness ratio of the member. This is required to prevent undesirable lateral movement or excessive vibration. A slenderness limits specified in IS: 800-2007 for tension members are given in table.

Table: Maximum values of effective slenderness ratio as per IS: 800-2007:

Member	Maximum effective slenderness ratio (l/r)
A tension member in which a reversal of direct stress occurs due to loads other than wind or seismic (inertia) forces.	180
A member subjected to compressive forces resulting only from a combination of wind/earthquake actions, provided the deformation of such a member does not adversely affect the stresses in any part of the structure.	250
A member normally acting as a tie in a roof truss or a bracing member which is not considered effective when subject to reversal of stress resulting from the action of wind or earthquake forces.	350
Members always in tension (other than pre-tensioned members).	400

2.1.3 Design of Tension Members and Gusset Plate

The design of a tension member involves selection of the type and the size of the section such that it can carry the computed factored design tensile load satisfactorily. The type of member is usually dictated by the location where the member is to be used. For example, in the case of roof trusses, angles or pipes are generally used.

Depending upon the span of the truss, the location of the member in the truss and the force in the member, either single angle or double angles may be used. Single angle is common as the web members of a roof truss and the double angles are common in rafter and tie members. The sizing process consists in determining cross sectional geometries of the members with minimum sectional area which satisfy the performance criteria stipulated in codes.

The performance criteria generally include design strength requirements, stiffness or stability and serviceability requirements. Thus, design process for a tension member goes hand in hand with design of the end connection.

The Design Strength Requirements Require the Strength Computations for,

- Gross section yielding.
- Tensile rupture.
- Shearing of bolts.
- Bearing of bolts.
- Block shear.

The section selected for the member should be the lightest steel section with design strength T_d greater than or equal to the maximum design tension load T_u acting on it. The sizing of steel member is essentially a trial and modification or iterative process involving selection of a trial section and analysis of its capacity, although sometimes tables available in steel manuals enable direct selection of desirable section.

For a design of tension member area required is estimated and a (lightest) section is selected from the steel section handbook providing the corresponding area and the section is checked for its suitability. In the next trial, it may be necessary to consider a slightly larger or smaller section and repeat the checking process.

The design process is repeated till a (lightest) cross section of such a size is obtained that satisfies the strength and stiffness requirements stipulated in the Code, there may be some rounding up or down in the process of selecting the final section.

The major steps involved in the design process are:

1. Calculation of Design or Factored Load

The factored load is given by,

$$T_u = \gamma_{DL} \times DL + \gamma_{LL} \times LL \qquad \text{...(1)}$$

As per IS: 800, $\gamma_{DL} = \gamma_{LL} = 1.5$ and thus $T_u = 1.5\,(DL + LL)$

2. Estimation of Area Required

At the first instance, the net effective area required is calculated from the design tension and the ultimate strength of the material as follows,

$$A_n = T_u / \left(0.9 f_u / \gamma_{ml}\right) \qquad \text{...(2)}$$

From the net-sectional area calculated above, the gross area required is estimated, allowing for some predetermined number and size of bolt/rivet holes in the section or the assumed efficiency index in the case of angles and threaded rods. The required gross area is then checked against that required from the yield strength criterion which is given by,

$$A_g = T_u / \left(f_y / \gamma_{mo} \right)$$

...(3)

Based on the estimated gross area, a suitable trial section is selected from the steel section tables. The number and arrangement for bolt/rivet holes are decided appropriately and the trial member is analyzed to evaluate the actual design strength as explained in the foregoing sections. If the actual design strength is smaller than or too large compared to the design force, a new trial section is selected and the analysis is repeated until a satisfactory design is obtained.

3. Stiffness Requirement

The tension members, in addition to meeting the design strength requirements, frequently have to be checked for their adequacy in stiffness or stability. This is done to ensure that the member does not sag too much during service due to self-weight or the eccentricity of end plate connections.

Following limitations on the slenderness ratio, L_e/r of members subjected to tension have been stipulated the IS: 800:

- In the case of members that are normally under tension but may experience compression due to stress reversal caused by wind/earthquake loading $L_e/r \le 250$.

- In the case of members that are designed for tension but may experience stress reversal for which it is not designed (as in X bracings) $L_e / r \le 350$.

- In the case of members subjected to tension only $L_e / r \le 400$.

Where, L_e and r, are the effective length of the member and minimum radius of gyration of the cross section, respectively. In the case of rods used as a tension member in X-bracings, the slenderness ratio limitation need not be checked, if the rods are pre-tensioned by using a turnbuckle or other such mechanical arrangement.

4. Design of Rolled Steel Sections

The shapes of sections selected depend on the types of members which are fabricated and to some extent on the process of erection. Many steel sections are readily available in the market and have frequent demand. Such steel sections are known as regular steel sections.

Some steel sections are rarely used. Such sections are produced on special requisition and are known as special sections. 'ISI Handbook for Structural Engineers' gives nominal dimensions, weight and geometrical properties of various rolled structural steel sections.

Gusset Plates

Gusset plates are thick sheets of steel that are used to connect beams and girders to columns or to connect truss members. A gusset plate can be fastened to a permanent member either by bolts, rivets or welding or a combination of the three. Gusset plates not only serve as a method of joining steel members together but they also strengthen the joint. They are used in bridges and buildings as well as other structures.

Problems

1. Let us design a double angle tension member connected on each side of a 10 mm thick gusset plate, carry an axial factored load of 375 kN. Use 20 mm black bolts. Assume shop connection.

Solution:

Given:

   Axial factored load of 375 kN.

   20 mm black bolts.

   Thickness -10 mm

Area required from the consideration of yielding,

$$= \frac{1.1 \times 375 \times 1000}{250} = 1650 \, \text{mm}^2$$

Try 2 ISA 7550, 8 mm thick which has gross area,

   $= 2 \times 938 = 1876 \, \text{mm}^2$

(i) In double shear $= \left[\frac{\pi}{4} \times 20^2 + 0.78 \times \frac{\pi}{4} \times 20^2 \right] \times \frac{400}{\sqrt{3}} \frac{1}{1.25}$

(ii) Strength in bearing:

   Taking e = 40 mm, p = 60 mm,

   K_b is smaller of $\frac{40}{3 \times 22}, \frac{60}{3 \times 22} - 0.25, \frac{400}{410}, 1.0$

i.e.,

   $K_b = 0.606$

$$V_{dsb} = \frac{1}{1.25} \times 2.5 \times 6.606 \times 20 \times 8 \times 400 = 77568 \text{ N}$$

Bolt value = 77568 N

Number of bolts required $= \dfrac{375000}{77568} = 4.83$

Provide 5 bolts in a row as shown in figure.

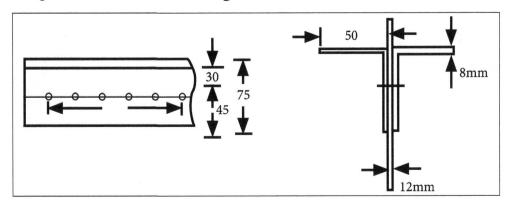

Checking the Design

(i) Strength against yielding,

$$\frac{A_g f_y}{\gamma_{mo}} = \frac{1876 \times 250}{1.1} = 426364 \text{ N} > 375 \times 1000$$

Hence, the design is safe.

(ii) Strength of plate in rupture:

Area of connected leg,

$$A_{nc} = 2\left(75 - 22 - \frac{8}{2}\right) \times 8 = 784 \text{ mm}^2$$

Area of outstanding leg,

$$A_{gc} = 2 \times \left(50 - \frac{8}{2}\right) \times 736 \text{ mm}^2$$

$$\beta = 1.4 - 0.076 \times \frac{w}{t} \times \frac{f_y}{f_y} \times \frac{b_s}{l_c}$$

$$= 1.4 - 0.076 \times \frac{50}{8} \times \frac{250}{410} \times \frac{77}{240}$$

$$= 1.307$$

$$\therefore T_{dn} = \frac{0.9 f_u A_{nc}}{\gamma_{ml}} + \beta \frac{Ag_o f_y}{\gamma_{mo}}$$

$$= \frac{0.9 \times 410 \times 784}{1.25} + 1.307$$

$$= \frac{736 \times 250}{1.1} = 450062 > 175000\,\mathrm{H}$$

Hence, the design is safe.

(iii) Strength against block shear failure:

Per angle,

$A_{vg} = (40 + 60 \times 4) \times 8 = 2240$ mm²

$A_{vn} = (40 + 60 \times 4 - 4.5 \times 22) \times 8 = 1448$ mm²

$A_{tg} = (75 - 35) \times 8 = 320$ mm²

$A_{in} = (75 - 35 - 0.5 \times 22) \times 8 = 232$ mm².

2. Let us determine the tensile load carrying capacity of 2ISA 75 × 75 × 8mm placed back to back of a 10mm thick gusset plate using a single row of 4 nos. 16mm diameter bolts at distance of 40mm from the toe of the angle. Take pitch = 50mm and end distance = 30mm. The length of the member is 4m. Use Fe410 grade steel.

Solution:

Given:

Dia =16mm

2ISA 75 x75 x8mm

ω =75 mm

ω_t = ?

Thick (t) = 8mm

Tensile Load Carrying Capacity

2 ISA → 75 × 75 × 8mm

Connected to 10mm gusset plate using single row of 4 Nos. of ϕ 16mm at a distance of 40mm from toe of angle,

Pitch = 50 mm

End distance = 30mm

Length of member = 4 m

Use F3410 grade steel.

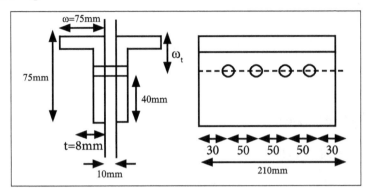

Gross Area of Angles = 2 × 1138 = 2276 mm²

(i) Strength Against Yielding:

$$T_{dg} = \frac{A_g f_y}{\gamma mo} = \frac{2276 \times 250}{1.1} = 517272\,N$$

(ii) Strength Against Rupture:

Net area connected leg,

$$A_{nc} = \left[75 - \frac{08}{2} - 18\right] \times 8 \times 2$$

$$= 7848 \text{ mm}^2$$

Gross area outstanding leg, $A_{ng} = 2\left[75 - \frac{8}{2}\right]8 = 1136\,mm^2$

$b_s = [\omega + \omega_t \times t] = 75 + 35 - 8$

$b_s = 102$ mm

$L_c = 3 \times 50 = 150$mm

$$\beta = 1.4 - 0.076\left[w/t\right]\left[b_s/L_c\right] \leq \left[\frac{f_u \gamma_{mo}}{f_y \gamma_{m1}}\right] \geq 0.7$$

[as per Cl. 6.33 of IS 800:2007].

$$= 1.4 - 0.076[75/8]\left[\frac{250}{410}\right]\left[\frac{102}{150}\right]$$

$\beta = 1.105$

$$T_{dn} = \frac{0.9 \times A_{nc} \times f_u}{\gamma_{m1}} + \frac{\beta A_{go} f_y}{\gamma_{mo}}$$

$$= \frac{0.9 \times 9848 \times 410}{1.25} + \frac{1.105 \times 1136 \times 250}{1.1}$$

$T_{dh} = 575$

(iii) Strength Against Block Shear:

$A_{vg} = [[30 + [3 \times 50]]] \times 8 = 1440 \text{ mm}^2$

$A_{un} = [30 + (3 \times 50) - 3.5 \times 18] \times 8 = 936 \text{mm}^2$

$A_{tg} = [75 - 35] [8] = 3200 \text{ mm}^2$

$A_{tn} = [75 - 35 - (0.5 \times 18)] [8] = 248 \text{ mm}^2$

$$T_{db_1} = \frac{A_{vg} f_y}{\sqrt{3} \gamma_{mo}} + \frac{0.9 A_{tn} f_u}{\gamma_{m1}}$$

$$= \frac{1440 \times 250}{\sqrt{3} \times 1.1} + \frac{0.9 \times 248 \times 410}{1.25}$$

$T_{db_1} = 262.16 \text{kN}$

$$T_{db_2} = \frac{0.9 A_{vn} f_u}{\sqrt{3} \gamma_{m_1}} + \frac{0.9 A_{tg} f_y}{\gamma_{m_0}}$$

$$= \frac{0.9 \times 936 \times 410}{\sqrt{3} \times 1.125} + \frac{320 \times 250}{1.1}$$

$T_{db_2} = 232.25 \text{kN}$

T_{db} for single angle = 232.25 kN

T_{db} for 2 angle = 2 × 232.25

$\quad$ = 464.5 kN

∴ The design Tensile Load carrying is, 464.5 kN.

3. Let us design a tension member to carry a factored tensile load of 300 kN. The 3m long tension member is connected to a gusset plate 16mm thick with one line of 20mm diameter bolts of grade 4.6. Use Fe410 grade steel.

Solution:

Given:

Tensile load of 300 kN

Plate Thickness = 16mm

Bolt dia=20 mm

Design Tension Member

Factored tensile Load = 300 kN

Length of member = 3 m

Gusset plate = 16mm

Diameter of bolts = 20 mm

Grade of bolts = 4.6

Use Fe410 grade steel,

$$T_u = \frac{A_g f_y}{\gamma_{mo}}$$

$$A_g \text{ [required]} = \frac{T_u \gamma_{mo}}{fy} = \frac{300 \times 10^3 \times 1.1}{250}$$

$$A_g = 1320 \text{ mm}^2$$

Consider 2 Angles with long leg outstanding and referring Steel Table, let us choose two Angles ISA 80 × 50 × 6 mm with a gross Area = 1492mm².

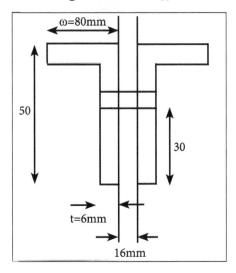

1. Strength in Double Shear

$$= \frac{f_{ub}}{\sqrt{3}} \frac{1}{\gamma_{m_0}} \left[\frac{\pi}{4} [0.78] d^2 + \frac{\pi}{4} d^2 \right]$$

$$= \frac{400}{\sqrt{3}} \times \frac{1}{1.25} \left[\frac{\pi}{4} 0.78 \times 20^2 + \frac{\pi}{4} 20^2 \right]$$

$$= \frac{400}{\sqrt{3}} \times \frac{1}{1.25} \left[\frac{\pi}{4} 0.78 \times 20^2 + \frac{\pi}{4} 20^2 \right]$$

$$= 103265 \, N$$

2. Strength in Bearing, V_{dpt}

Assume e = 30 mm; P = 50 mm

$$K_b \text{ is least of } \frac{e}{3d_0}, \frac{P}{3d_0} - 0.25, \frac{f_{yt}}{f_u}, 1$$

$$= \frac{30}{3 \times 22}, \frac{50}{3 \times 22} - 0.25, \frac{400}{410}, 1$$

$$= 0.4545, \, 0.5076, \, 0.9756, \, 1$$

$$K_b = 0.4545$$

$$V_{dpb} = \frac{2.5 \times K_b \, dt \, f_u}{\gamma_{m_0}}$$

$$= \frac{2.5 \times 0.4545 \times 20 \times 6 \times 410}{1.25} = 44722 \, N$$

Bolt Value = 44722 N

No. of bolt required = $\dfrac{300000}{44722} = 6.7 \approx 7 \, \text{Nos.}$

3. Checking the Design

i. Strength Against Yielding

$$T_{dg} = \frac{A_g f_y}{\gamma m_1} = \frac{1492 \times 250}{1.1}$$

$$T_{dg} = 339091 \, N > 300000 \, N$$

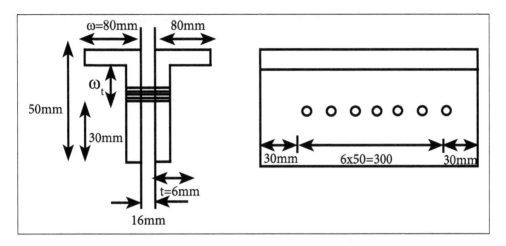

ii. Strength Against Rupture

Net area of connected leg,

$$A_{nc} = 2\left[50 - \frac{6}{2} - 22\right]6 = 300\,mm^2$$

Gross area of outstanding leg,

$$A_{gc} = 2\left[80 - \frac{6}{2}\right] \times 6 \times 924\,mm^2$$

$\omega = 80$ mm, t = 6 mm, $b = [80 + 20 - 6] = 94$ mm

$L_c = 6 \times 50 = 300$ mm

$$\beta = 1.4 - 0.076\left[\frac{80}{6}\right]\left[\frac{250}{410}\right]\left[\frac{94}{300}\right] \le 1.4432 \ge 0.7$$

$$= 1.2064 \le 1.4432 \ge 0.7$$

$$\beta = 1.2064$$

$$T_{dn} = \frac{0.9[410][300]}{1.25} + \frac{1.2064 \times 924 \times 250}{1.1}$$

$$= 341904\,N > 300000N$$

iii. Strength Against Block Shear

$$A_{vg} = [30 + (6 \times 50)] \times 6 = 1980\,mm^2$$

$$A_{vn} = [30 + (6 \times 50) - 6.5 \times 22][6] = 1122\,mm^2$$

$$A_{tg} = [50 - 20]6 = 180\,mm^2$$

$A_{tn} = [50 - 20 - 0.5 \times 22] \times 6 = 114 \text{ mm}^2$

$T_{db_1} = \dfrac{A_{vg} f_y}{\sqrt{3}\, \gamma_{m_o}} + \dfrac{0.9 A_{tn} f_u}{\gamma_{m_1}}$

$T_{db_1} = \dfrac{1980}{\sqrt{3}} \times \dfrac{[250]}{1.1} + \dfrac{0.9 \times 114 \times 410}{1.25}$

$T_{db1} = 293468 \text{ N}$

$T_{db_2} = \dfrac{A_{vn} f_u \times 0.9}{\sqrt{3} \times \gamma_{m_1}} + \dfrac{A_{tg} f_y}{\gamma_{m_o}}$

$\qquad = \dfrac{0.9 \times 1122 \times 410}{\sqrt{3} \times 1.25} + \dfrac{180 \times 250}{1.1}$

$T_{db_2} = 232141 \text{ N}$

T_{db} form Single Angle = 232141 for Double Angle = 2 × 232141

$T_{db} = 464282 \text{ N} > 300000 \text{ N}$

Hence the design is safe.

4. A double angle ISA 75 MM × 75 MM × 8 mm back to back welded to one side 12 mm gusset have allowable stress 150 MPa. Let us determine allowable tensile load on the member and weld length and overlap length of gusset plate.

Solution:

Given:

Angle = ISA 75 MM × 75 MM × 8 mm

Allowable stress = 150 MPa

Gross area of angle ISA75 × 75 × 8 mm

= 2 × 1138 = 2276 mm²

$A_1 = \dfrac{A}{2} = \dfrac{1138}{2} = 569 \text{ mm}^2$

Effective area,

$k = \dfrac{A_1}{A_1 + \dfrac{A_2}{5}} = \dfrac{569}{569 + \dfrac{569}{5}}$

$\quad = 0.833$

Effective area,

$$= 2 (A_1 + KA_2)$$

$$= 2[569 + 0.833 (569)$$

$$= 2086 \text{ mm}^2$$

Allowable stress in tension = 150 MPa.

Maximum allowable tensile load = Effective area × Permissible stress,

$$= 2086 \times 150$$

$$= 312900 \text{ N}$$

$$= 312.9 \text{ kN}$$

Thickness of angle t = 8 mm

Size of Thickness of Well

(i) Maximum size of weld $= \dfrac{3}{4}$ (thickness of angle)

$$= \dfrac{3}{4} (8) = 6 \text{ mm}$$

Thickness of thin plate 8– 1.5 mm,

$$= 8 - 1.5 = 6.5 \text{ mm}$$

(ii) Maximum size = 3 m

$$= 0.7 \text{ (Size of weld)}$$

$$= 0.7(5) = 3.5 \text{ mm}$$

$$f_{wd} = \frac{f_{wn}}{\gamma_{mw}} = \frac{f_u/\sqrt{3}}{\gamma_{mw}}$$

Assume,

$$f_u = 410 \text{ MPa}, \gamma_{mw} = 1.25$$

$$f_{wd} = \frac{410\sqrt{3}}{1.25} = 189.39 \text{ MPa}$$

$$\therefore \text{ Strength of weld} = \frac{1(0.7 \times 5)}{1.25}\left(\frac{410}{\sqrt{3}}\right) = 662.82\,\text{N} \approx 662\,\text{N}$$

$$= \frac{\text{force}}{\text{Strength}}$$

$\therefore$ Length of weld required $= \dfrac{312900}{662} = 472.66 \, \text{mm} = 480 \, \text{mm}$.

$\therefore$ Length of weld $= \dfrac{486}{2} = 240 \, \text{mm}$

Overlap length required,

$$= \frac{240 - 75}{2}$$

$$= 82.5 \text{ say } 85 \text{ mm}$$

5. An ISA 100 mm × 100 mm × 12 mm is used as a tin riveted to a gusset plate with 24 mm rivets arranged in one row along the length of the angle. Let us determine the allowable tension on the angle and the allowable tensile stress is 150 MPa.

Solution:

Given:

An ISA 100 mm × 100 mm × 12 mm

Diameter of the rivet hole = 24 + 1.5 = 25.5 mm

$$A_{net} = A_{stress} = A_1 + A_2 k$$
$$k = \frac{3A_1}{3A_1 + A_2}$$
$$A_1 = A_2 = \frac{A}{2}$$

Grass area 100 × 100 × 12 mm = 2259 mm².

$$A_1 = A_2 = \frac{2259}{2} = 1129.5 \text{ mm}^2$$

$$k = \frac{3(1129.5)}{3(1129.5) + 1129.5} = 0.75$$

$$A_{net} = A_{stress} = A_1 + A_2 k$$

$$= 1129.5 + 1129.5 \, (0.75)$$

$$= 1976.625 \text{ mm}^2$$

Allowable tension = Allowable Tensile Stress × A$_{stress}$

$\qquad$ = 150 (1976.625)

$\qquad$ = 296493 N ≈ 296.493 kN

6. An angle section 500 × 30 × 6 mm is used as a tension member with its longer leg connected by 12 mm diameter rivets. Let us calculate its strength and take permissible stress in axial tension as l50N/mm².

Solution:

Given:

$\qquad$ Angle = 500 × 30 × 6 mm

$\qquad$ Longer leg connected by = 12 mm diameter rivets.

$\qquad$ Permissible axial stress in axial tension = 150 N/mm²

Rivet

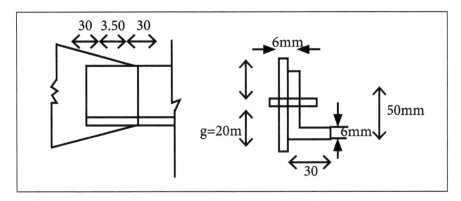

Assume, g = 20mm

(i) Strength Governed by Yielding:

$$T_{dg} = \frac{A_g f_y}{\gamma_{m_o}}$$

$$A_g = \left[50 - \frac{6}{2} + 30 - 6/2\right] \times 6$$

$$= 444 \text{ mm}^2$$

$$T_{dg} = \frac{0444 \times 250}{1.1} = 100909.0 \, N$$

(ii) Strength Governed by Tearing at Critical Sections:

Refer to IS 800-2007.

$$T_{dn} = \left[\frac{0.9 \times A_{nc} f_u}{\gamma_{m_1}}\right] + \left[\frac{\beta A_{go} f_y}{\gamma_{m_o}}\right]$$

$$A_{nc} = \left[50 - \frac{6}{2} - 12\right] 6 = 210 \, mm^2$$

$$A_{go} = [30 - 6/2]6 = 162 mm^2$$

$$\beta = 1.4 - 0.076 \left[\frac{w}{t}\right]\left[\frac{f_y}{f_u}\right]\left[\frac{b_s}{L_c}\right] \le \frac{f_u \gamma_{m_o}}{f_y \gamma_{m_1}} \ge 0.7$$

$\omega = 30$ mm

$\omega_t = 20$ mm ; t = 6 mm

$b_s = 30 + 20 - 6 = 44$ mm

$L_c = 3 \times 50 = 150$ mm

$$\beta = 1.4 - 0.076\left[\frac{30}{20}\right]\left[\frac{250}{150}\right]\left[\frac{44}{150}\right] = \frac{440 \times 1.1}{250 \times 1.25} \ge 0.7$$

$$= 1.34 \le 1.443 \ge 0.7$$

$$T_{dn} = \frac{0.9 \times 210 \times 410}{1.25} + \frac{1.34 \times 162 \times 250}{1.1}\beta = 1.34$$

$$= 61992 + 49336 = 111328 \, N$$

(iii) Block Shear Strength:

$$T_{dn} = \frac{A_{vg} f_y}{\sqrt{3}\,\gamma_{m_o}} + \frac{0.9 A_{tn} f_u}{\gamma_{m_1}}$$

$A_{vg} = [30 + [3 \times 50]]\,6 = 1080 \, mm^2$

$A_{vn} = [30 + (3 \times 50) - 3 - 5[12]] \times 6 = 828 \, mm^2$

$A_{fg} = [50 - 20] \times 6 = 180 \, mm^2$

$A_{tn} = [50 - 20 - 0.5 \times 12]6 = 144 \, mm^2$

$$T_{db_1} = \frac{1080 \times 250}{\sqrt{3} \times 1.1} + \frac{0.9 \times 144 \times 490}{1.25} = 184222 \, N$$

$$T_{db_2} = \frac{A_{tg}f_y}{\gamma_{m_0}} + \frac{0.9 A_{vn}f_u}{\sqrt{3}\,\gamma_{m_1}}$$

$$= \frac{180 \times 250}{1.1} + \frac{0.9 \times 828 \times 410}{\sqrt{3} \times 1.25} = 182028\,N$$

$$T_{db} = 182028\ N$$

Strength of member = 100909N

If Welds,

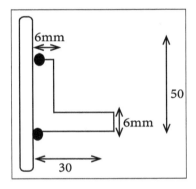

$$A_g = 447\,mm^2$$

$$A_{nc} = \left[50 - \frac{6}{2}\right]6 = 282\,mm^2$$

$$A_{nc} = \left[30 - \frac{6}{2}\right]6 = 162\,mm^2$$

(iv) Strength Governed by Yielding:

$$T_{dg} = \frac{A_g f_y}{\gamma_{m_0}} = \frac{447 \times 250}{1.1} = 101590\,N$$

(v) Strength of Plate from Rupture:

$$T_{dn} = \frac{0.9 f_u A_{nc}}{\gamma_{m_1}} = \frac{\beta A_{go} f_y}{\gamma_{m_0}}$$

$$\beta = 1.4 - 0.076 \left[\frac{w}{t}\right]\left[\frac{f_y}{f_u}\right]\left[\frac{b_s}{L_c}\right] \leq \frac{f_u \gamma_{m_0}}{f_y \gamma_{m_1}} \geq 0$$

$\omega = 30$ mm, t = 6 mm, $b_s = 30$ mm, $L_c = [210]$ mm

$$\beta = 1.4 - 0.076 \left[\frac{30}{6} \right] \left[\frac{250}{210} \right] \leq \left[\frac{410}{250} \times \frac{1.1}{1.25} \right] \geq 0$$

$$= 1.36 \leq 1.44 \geq 0.7\beta = 1.36$$

$$T_{dn} = \frac{0.9 \times 410 \times 282}{1.25} + \frac{1.36 \times 162 \times 250}{1.1} = 133319 \, N$$

(vi) Block Shear Strength:

Assume gusset plate = 10 mm

$$A_{vg} = 2 \left[\frac{210 + 210}{2} \right] \times 10 = 4200 \, mm^2$$

$$A_{vg} = A_{vn} = 4200 \, mm^2$$

$$A_{vg} = A_{tn} = 50 \times 10 = 500 \, mm^2$$

$$T_{bd_1} = \frac{4200 \times 250}{\sqrt{3} \times 1.1} + \frac{0.9 \times 500 \times 410}{1.25}$$

$$T_{db_1} = 698707 \, N.$$

$$T_{bd_2} = \frac{0.9 \times 4200 \times 410}{\sqrt{3} \times 1.25} + \frac{500 \times 250}{1.1}$$

$$T_{db_2} = 829458 \, N$$

$$T_{db} = 698707 \, N$$

Strength of tension member = 101590 N

7. Let us design a tension member of length 3.6 m between c/c of intersections and carrying a pull of 150 kN. The member is subjected to reversal of stresses.

Solution:

Given:

Length = 3.6 m

Gross area of the angle required.

$$T_u = \frac{A_g f_u}{\gamma_{mo}}$$

$$A_g = \frac{150 \times 1000 \times 1.1}{250} = 880 \, mm^2 = 880 \, mm^2$$

Use ISA 100 65, 8 mm with gross area A_g = 1257 mm².

Let us use 10 mm gusset plate for connection.

Diameter of bolt = 20 mm

Diameter of bolt hole = 22 mm.

Strength of one bolt in single shear

$$V_{dsb} = \frac{f_{ub}}{\sqrt{3}}\left(\frac{n_n A_{nb} + n_s^{\cdot} A_{sb}}{\gamma_{mb}}\right)$$

Where,

$$= \frac{400}{\sqrt{3}}\left(\frac{1\times0.78\times\pi\times\dfrac{20^2}{4}}{1.25}\right) = 45251\ N = 45.25\ kN$$

Adopt edge distances e = 40 mm, pitch P = 60 mm,

$k_b = 0.6061$

$$V_{dpb} = \frac{2.5\ k_b\ dt\ fu}{\gamma_{mb}}$$

$$= \frac{2.5(0.6061)(20)(8)(410)}{1.25} = 79520\ N = 79.52\ kN$$

Bolt Value = 45251 N = 45.251 kN

Number of bolts = $\dfrac{(150)(1000)}{45251}$ = 5 bolts

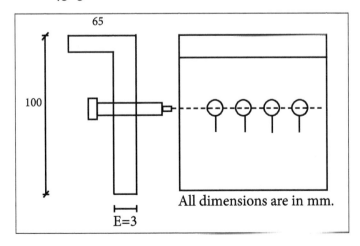

All dimensions are in mm.

Checking the Design

(i) Strength from yielding consideration,

$$= \frac{A_g f_y}{\gamma_{mo}} = \frac{1257(250)}{1.1}$$

$$= 285.682 \text{ N} > 20000 \text{ N}$$

(ii) Strength from rupture of critical section,

Net area of connected leg:

$$A_{nc} = (100 - 22 - 8/2) \times 8 = 592 \text{ mm}^2$$

Gross area of outstanding leg,

$$= \left(65 - \frac{8}{2}\right) \times 8 = 488 \text{ mm}^2$$

To calculate β,

$$W = 65 \text{ mm}, t = 8 \text{ mm}, b_s = (65 + 40 - 8) = 97 \text{ mm}$$

$$L_C \Rightarrow \text{Length of member between outermost.}$$

Fasteners = 4(60) = 240 mm,

$$\beta = 1.4 - 0.076 \left(\frac{w}{t}\right)\left(\frac{f_y}{f_u}\right)(bs_{lc}) \leq \frac{fu}{fy}\frac{\gamma_{m_o}}{\gamma_{me}} \geq 0$$

$$\beta = 1.4 - 0.076\left(\frac{65}{8}\right)\left(\frac{250}{410}\right)\left(\frac{97}{240}\right)$$

$$= 1.243 \leq 1.4433 \geq 0.7$$
$$\beta = 1.248$$

$$T_{bn} = \frac{0.9 \, fu \, A_{nc}}{\gamma_{x\lambda}} + \frac{\beta \, A_{go} \, fy}{\gamma_{mo}}$$

$$= \frac{0.9(410)(592)}{1.25} + \frac{1.248(488)(25)}{1.1}$$

$$= 174750 + 138415$$

$$= 313173 \text{ N} > 200000 \text{ N}$$

(iii) From Block Shear Consideration:

$$A_{vg} = [40 + 60(4)] \, (8) = 2240 \text{ mm}^2$$

$$A_{vn} = [40 + 60(4) - 4.5(22)] \, (8) = 1448 \text{ mm}^2$$

$$A_{fg} = [100 - 40] \, (8) = 480 \text{ mm}^2$$

$$A_{tn} = [100 - 40 - 0.5(22)] \, (8) = 392 \text{ mm}^2$$

$$T_{db1} = \frac{A_{vg} \, fy}{\sqrt{3} \, \gamma_{mo}} = \frac{0.9 \, A_{tn} \, fy}{\gamma_{me}}$$

$$= \frac{2240(250)}{\sqrt{3}(1.1)} + \frac{0.9(392)(410)}{1.25} = 409642 \text{ N}$$

$$T_{db2} = \frac{0.9 A_{vn} \, fu}{\sqrt{3} \, \gamma_{mo}} = \frac{A_{vg} \, fy}{\gamma_{mo}}$$

$$= \frac{0.9(1498)(410)}{\sqrt{3}(1.25)} + \frac{480(250)}{1.1} = 355886 \text{ N}$$

$$T_{db} = 355886 \text{ N} > 200000 \text{ N}$$

Hence the design is safe.

Check for maximum 1/r value.

Referring to steel tables γ_{min} for angle section,

$$100 \times 65 \times 8 \text{ mm} = 13.9 \text{ mm}$$

$$\frac{1}{\gamma_{mim}} = \frac{3000}{13.9} = 215.8 < 350$$

8. Let us design a tension member to carry a load 300 kN. The two angle placed back to long leg outstanding are desirable. The length of the member is 3 m.

Solution:

Given

Load = 300 kN

Length = 3m

Area required from yielding consideration.

$$T_u = \frac{A_g \, fy}{\gamma_{mo}} \Rightarrow A_{g \text{ required}} \frac{T_u \, \gamma_{mo}}{f_y}$$

$$A_{g\,required} = \frac{300(1000)(1.1)}{250} = 1320\,mm^2$$

(i) In Double Shear:

$$= \frac{400}{\sqrt{3}} \cdot \frac{1}{1.25}\left[\frac{\pi}{4}(0.78)\times 20^2 + \frac{\pi}{4}\times 20^2\right] = 103365\,N$$

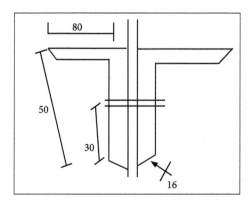

(ii) Strength in Bearing:

Take e = 30 mm, P = 50 mm

$$k_b \text{ is least of } \frac{30}{3(22)}; \frac{50}{3(22)}-0.25; \frac{400}{410}; 1.0$$

least of 0.4545; 0.5076; 0.9756; 10

$$k_b = 0.4545$$

Strength in Bearing,

$$V_{dpb} = \frac{2.5\,k_b\,dt\,fu}{\gamma_{mb}}$$

$$= \frac{2.5(0.4545)(20)(6)(410)}{1.25} = 44722\,N$$

∴ Bolt Value = 44722 N

$$\text{No. of bolts Required} = \frac{300000}{44722} = 6.7 \approx 7\,bolts$$

(iii) Strength against design:

$$= \frac{A_g \cdot fy}{1.1} = \frac{1492(250)}{1.1}$$

$$= 33909\,N > 300000\,N$$

(iv) Strength against rupture:

Net area of connected,

$$A_{nc} = 2\left(50 - \frac{6}{2} - 22\right)(6) = 300\,\text{mm}^2$$

$$A_{go} = 2\left(20 - \frac{6}{2}\right)(6)\,(2) = 924\,\text{mm}^2$$

W = 80 mm, t = 6 mm

$b_s = (80 + 20 - 6) = 94$ mm, $L_C = 6(50) = 300$ mm

$$\beta = 1.4 - 0.076\left(\frac{80}{6}\right)\left(\frac{25}{410}\right)\left(\frac{94}{300}\right)$$

$\beta = 1.2064$

$$T_{du} = \frac{0.9(410)(300)}{1.25} + \frac{1.2064(924)(250)}{1.1}$$

$$= 3419.4 > 300000\,\text{N}.$$

(v) Strength against Block Shear:

$$A_{vg} = [30 + (6 \times 50)]\,(6) = 1980\,\text{mm}^2$$

$$A_{vn} = [30 + 6 \times 50 - 6.5(22)]\,(6) = 1122\,\text{mm}^2$$

$$A_{tg} = [50 - 20]\,(6) = 180\,\text{mm}^2$$

$$A_{tn} = [50 - 20 - 05\,(22)]\,(6) = 114\,\text{mm}^2$$

Least Value T_{db1} and T_{db2} where

$$T_{db1} = \frac{A_{vg}\,f_y}{\sqrt{\gamma_{mo}}} = \frac{0.9\,A_{tn}\,fu}{\gamma_{m1}}$$

$$= \frac{1980(250)}{\sqrt{3}(1.1)} + \frac{0.9(114)(410)}{1.25} = 293468\,\text{N}$$

$$T_{db2} = \frac{0.9(1122)(410)}{\sqrt{3}(1.25)} + \frac{180(250)}{1.1} = 237141\,\text{N}$$

T_{db} for single angle = 232141 N

T_{db} for 2 angle = 2 × 232141 = 464782 N

C = 464282 N > 300000 N

Check for maximum $\dfrac{1}{\gamma_{min}}$ value $= \dfrac{2900}{13.9} = 208 < 350$

9. Let us design a tension member to carry a load of 300 kN. The two angles placed back to back with long leg outstanding are desirable. The length of the member is 2.9 m.

Solution:

Given:

Load = 300 kN

Length = 2.9m

Met over required,

$$\dfrac{180 \times 10^3}{150} = 1200\,mm^2$$

Let us try a single equal angle 90 mm × 90 mm × 90 mm. Net area of the connect leg,

$A_1 = (90 - 5)10 - 21.5 \times 10 = 635\ mm^2$

Area of the outstanding leg,

$A_2 = (90 - 5)10 = 850\ mm^2$

$k = \dfrac{3A_1}{3A_1 + A_2} = \dfrac{3 \times 635}{3 \times 635 + 850} = 0.6915$

Effective area,

$A_{eff} = A_1 + kA_2 = 635 + (0.6915 \times 850)$

$A_{eff} = 1222.775\ mm^2$

Safe tension of the member $= \dfrac{1222.775 \times 150}{1000} = 183.42\,kN$

But tension in the member is 180 kN. Therefore design is safe.

Let us try a single unequal angle.

100mm × 65 mm × 10 mm

$A_1 = (100 - 5)10 - 21.5 \times 10 = 735\ mm^2$

Area of the outstanding leg,

$A_2 = (65 - 5)10 = 600\ mm^2$

$$K = \frac{3A_1}{3A_1 + A_2} = \frac{3 \times 735}{3 \times 735 + 600} = 0.7861$$

Effective area of the member,

$$A_{eff} = A_1 + KA_2 = 735 + 0.7861 \times 600$$

$$= 1,206.66 \text{ mm}^2$$

Safe tension for the member $= \dfrac{1206.66 \times 150}{1000} = 180.999 \text{ kN}$

The design is safe.

10. Let us design a splice to connect a 300 × 20 mm plate with a 300 × 10 mm plate. The design load is 500 kN use 20 mm black bolts, fabricated in the shop.

Solution:

Given:

Design a splice to connect =300 × 20 mm and 300 × 10 mm

      Load =500 kN

Let double cover butt joint with 6 mm cover plates be used.

Strength of Bolts

Strength in double shear,

$$= \beta_{pk}\left(\frac{\pi}{4}d^2 + 0.78\frac{\pi}{4}d^2\right)\frac{f_c}{\sqrt{3} \times 1.25}$$

$$= 0.875 \times 1.78 \times \left(\frac{20}{4}\right) \times \frac{400}{\sqrt{3}} \times \frac{400}{\sqrt{3}} \times \frac{1}{1.25}$$

$$= 90400 \text{ N}$$

Strength in Bearing

Let edge distance = 40 mm and pitch 60 mm be used.

Then K_b is smaller of $\dfrac{e}{3d_o}, \dfrac{p}{3d_o} - 0.25, \dfrac{f_{ub}}{f_u}, 1.0,$

i.e., $\dfrac{40}{3 \times 22}, \dfrac{60}{3 \times 22} - 0.25, \dfrac{400}{410}, 1.0$

      $K_b = 0.606$

∴ Strength in bearing against 10 mm plate

$$= 2.5\ K_b\ dt\ f_u \times \frac{1}{\gamma_{wb}}$$

$$= 2.5 \times 0.606 \times 20 \times 10 \times 410 \times \frac{1}{1.25}$$

$$= 105897\ N$$

∴ Bolt value = 90400 N

Hence number of bolts required $= \dfrac{500\times1000}{990400} = 5.53$

Provide 6 bolts on each side of the joint as shown in figure.

Check for the Strength of Plate

(i) Strength against yielding of gross section:

$$= \frac{A_g f_y}{\gamma_{mo}} = \frac{300\times10\times250}{1.1}$$
$$= 681818\ N > 50000\ N$$

(ii) Strength against rupture:

$$A_n = (300 - 3 \times 22) \times 10 = 2340\ mm^2$$
$$\therefore T_{dm} = \frac{0.9\ A_n\ f_u}{\gamma_{ml}} = \frac{0.9\times234\times410}{1.25} = 690768\ N > 500000\ N$$

(iii) Block shear strength:

$$T_{db} = \frac{A_{vg}\ f_y}{\sqrt{3}\ \gamma_{mo}} + \frac{0.9\ A_{tn}\ f_u}{\gamma_{ml}}$$
$$= \frac{2000\times250}{\sqrt{3}\times1.1} + \frac{0.9\times1760\times410}{1.25}$$
$$T_{db} = 781984\ N$$

11. Let us design a tension splice to connect two plates of size 250 mm × 20 mm and 220 mm × 12 mm, for a design load of 250 kN.

Solution:

Given:

Plate size = 250 mm × 20 mm and 220 mm × 12 mm

Load =250 kN

Design Tension Splice:

250 × 20 mm and 220 × 12 mm.

Design Load = 250kN.

Let us provide double cover butt joint with 6 mm cover plates on either side with 20 mm diameter black belts.

Strength of Bolts

$d = 20$ mm, $d_{12} = 22$ mm, $\beta_{pk} = 1 - 0.0125\,[10] = 0.875$

Strength in Double Shear

$$= \beta_{PK}\left[0.78\frac{\pi d^2}{4} + \frac{\pi d^2}{4}\right]\frac{f_u}{\sqrt{3}} \times \frac{1}{1.25}$$

$$= 0.875\left[0.78 \times \frac{\pi \times 20^2}{4} + \frac{\pi \times 20^2}{4}\right] \times \frac{400}{\sqrt{3}} \times \frac{1}{1.25} = 90399\,\text{N}$$

Strength in Bearing

Let us provide edge distance, e = 40 mm

Pitch, P = 60 mm

$$k_b = \frac{40}{3 \times 22}; \frac{60}{3 \times 22} - 0.25; \frac{400}{410}; 1.0$$

$k_1 = n_o/n_f$

∴ Strength in bearing against 12 mm plate,

= 2.5 × 0.606 × 20 × 12 × 410 × 1/1.25

= 90399 N

Number of bolts required $= \dfrac{250 \times 1000}{90399} = 2.76 = 3\,\text{bolt}$

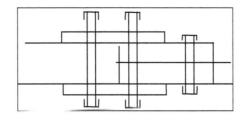

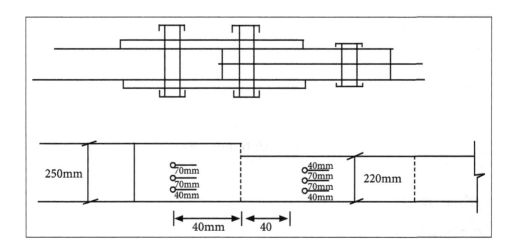

Checking the Design

(i) From Yielding Consideration:

Strength against yielding,

$$= \frac{[220\times12]\times250}{1.1}$$

$$= 5600000 > 250000\text{N}$$

(ii) From Rupture Consideration:

Strength against rupture,

$$T_{dn} = \frac{0.9\,A_n f_u}{\gamma_{m_1}}$$

$A_n = [220 - 3(22)] \times 12 = 1848 \text{ mm}^2$

$$T_{dn} = \frac{0.9\times1848\times410}{1.25} = 545529 > 250000\,\text{N}$$

(iii) Block Shear Strength:

$A_{vg} = 3 \times 40 \times 12 = 1440 \text{ mm}^2$

$A_{vn} = 3[[40] - 0.5 \times 22]\,12 = 1044 \text{ mm}^2$

$A_{kg} = [40 + 40]12 = 960 \text{ mm}^2$

$A_{tn} = [40 + 40 - 2 \times 22] \times 12 = 432 \text{ mm}^2$

$$T_{db_1} = \frac{A_{vg} f_y}{\sqrt{3} \cdot \gamma_{m_0}} + \frac{0.9 A_{tn} f_u}{\gamma_{m_1}}$$

$$= \frac{1440 \times 250}{\sqrt{3} \times 1.1} + \frac{0.9 \times 432 \times 410}{1.25}$$

$$T_{db_1} = 3.16477 \text{ N}$$

$$T_{db_2} = \frac{0.9 A_{vn} f_u}{\sqrt{3} \gamma_{m_1}} + \frac{A_{tg} f_y}{\gamma_{m_0}}$$

$$= \frac{0.9 \times 1044 \times 410}{\sqrt{3} \times 1.25} + \frac{960 \times 250}{1.1} = 39115 \text{ N}$$

$$T_{db} = 316477 \text{ N}$$

Let us provide 2 number of 30 mm dia meter half with one extra bolt in the caver.

2.2 Compression Members

A column and top chords of trusses, diagonals and bracing members are all examples of compression members. Columns are usually thought of as straight compression members whose lengths are considerably greater than their cross-sectional dimensions.

An initially straight strut or column, compressed by gradually increasing equal and opposite axial forces at the ends is considered first. Columns and struts are termed "long" or "short" depending on their proneness to buckling. If the strut is "short", the applied forces will cause a compressive strain, which results in the shortening of the strut in the direction of the applied forces. Buckling behaviour is thus characterized by large deformations developed in a direction (or plane) normal to that of the loading that produces it. When the applied loading is increased, the buckling deformation also increases. Buckling occurs mainly in members subjected to compressive forces.

If the member has high bending stiffness, its buckling resistance is high. Also, when the member length is increased, the buckling resistance is decreased. Thus the buckling resistance is high when the member is short or "stocky" conversely, the buckling resistance is low when the member is long or "slender".

Under incremental loading, this shortening continues until the column yields or "squashes". However, if the strut is "long", similar axial shortening is observed only at the initial stages of incremental loading. Thereafter, as the applied forces are increased in magnitude, the strut becomes "unstable" and develops a deformation in a direction normal to the loading axis and its axis is no longer straight, shown in figure. The strut is said to have "buckled".

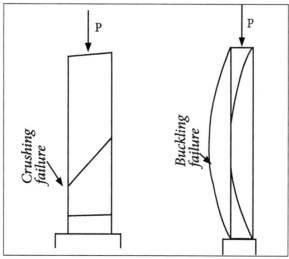

Short columnslong columns

Structural steel has high yield strength and ultimate strength compared with other construction materials. Hence compression members made of steel tend to be slender compared with reinforced concrete or prestressed concrete compression members. Buckling is of particular interest while employing slender steel members. Members fabricated from steel plating or sheeting and subjected to compressive stresses also experience local buckling of the plate elements.

Introduces a buckling in the context of axially compressed struts and identifies the factors governing the buckling behaviour. Both global and local buckling is instability phenomena and should be avoided by an adequate margin of safety.

Traditionally, the design of compression members was based on Euler analysis of ideal columns which gives an upper band to the buckling load. However, practical columns are far from ideal and buckle at much lower loads. The first significant step in the design procedures for such columns was the use of Perry Robertsons curves.

Modern codes advocate the use of multiple-column curves for design. Although these design procedures are more accurate in predicting the buckling load of practical columns.

2.2.1 Effective Length

Effective Length of Compression Members

The effective length KL is calculated from the actual length L of the member considering the rotational and relative translational boundary conditions at the ends. The actual length shall be taken as the length from centre-to-centre of its intersections with the supporting members in the plane of the buckling deformation.

In the case of a member with a free end the free standing length from the center of the intersecting member at the supported end shall be taken as the actual length.

Latticed Columns

Lattice columns are made of parallel supports interlaced with each other by strapping found in diamond or triangular patterns. The most famous example of lattice column is the Eiffel towers.

Eccentrically Loaded Column

Here, the load is transferred to the column from some eccentricity (the distance from the assumed point of application of the load to the centroid of column).

The columns are classified as under Euler's condition:

- Both end fixed.

- Both end pinned or hinged.

- One end fixed and other end hinged.

- One end fixed and other end free.

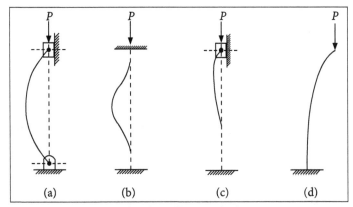

Types of end conditions of columns.

2.2.2 Slenderness Ratio

The slenderness ratio (λ) of a member is the ratio of the effective length to the appropriate radius of gyration ($\lambda = k_L/r$). This is valid only when the column has equal unbraced heights for both axis and end conditions are same for both axis. The appropriate radius of gyration is one which is minimum for a particular section.

For example a section asymmetrical about the centroidal axis angle section, figure (a) will

Analysis and Design of Steel Structures

bend about the principal axis V—V for which the radius of gyration is minimum. On the other hand, a section symmetrical about both the centroidal axis (I-Section) or even with one axis of symmetry (channel section, two angles back to back) will bend about one of the centroidal axis giving lesser radius of gyration. This is because for such sections the principal axis coincide with the centroidal axis. The slenderness ratios about XX and YY axis are,

$$\lambda_{xx} = l_{xx}/r_{xx} \quad r_{xx} = \sqrt{I_{xx}/A}$$

$$\lambda_{yy} = l_{yy}/r_{yy} \quad r_{yy} = \sqrt{I_{yy}/A}$$

And,

$$\lambda_{max} = l/r_{min} \quad r_{min} = \sqrt{I_{min}/A}$$

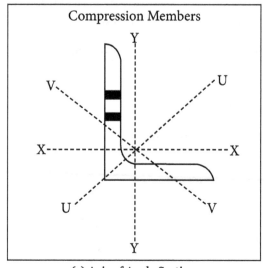

(1) Axis of Angle Section

To minimize steel requirements in column design, the slenderness ratio should be kept as small as possible. The designer can achieve this either by selecting a section which provides the largest value of the minimum radius of gyration without providing more area or by reducing the unsupported length of the column by some means. Maximum radius of gyration is obtained when material is farthest from the centroid. Thus, for a constant area the material gets thinner as the column size increases and in such a case local buckling limits the column size.

If feasible, the unsupported length can be reduced by furnishing intermediate supports to a compression member by attached construction and thus permitting use of a smaller section at high average stress. If this is possible in only one direction, then the value of the unsupported length will be different for the two directions. Sections with different radii of gyration in two directions may be so chosen as to obtain the values of the two slenderness ratios λ_{xx} and λ_{yy} with least difference.

The larger of the two would be used for the determination of the nominal axial compressive stress. It would be desirable to orient the section so that the axis having the smallest radius of gyration would be the one braced. Sometimes when the conditions at the column end, about the two axis are different, a balance between the two slenderness ratios (λ_{xx} and λ_{yy}) can be achieved by properly orienting the section and the material can be used to its maximum stress value.

An I-Section electric pole fixed at the base and carrying wires parallel to the road is an example wherein the wires provide a hinge condition in the direction of wires (l = 0.8L) and with free end condition in the perpendicular direction (l =2L.). The section is used to its full strength by providing flanges parallel to the direction of wires and thus attempting to equalize λ_{xx} and λ_{yy},

$$\lambda_{XX} = \frac{0.8L}{r_{YY}}, \lambda_{YY} = \frac{2L}{r_{XX}}$$

The slenderness ratio of compression members is limited because of the following reasons:

- The effect of accidental and construction (fabrication, transportation and erection) loads are automatically taken.

- The bracing members may be used as a walkway for workmen or to provide temporary support for equipments.

- To take care of the probability of members being subjected to unexpected vibrations.

The maximum permissible slenderness ratio for compression members are stated in table.

Table Maximum Slenderness Ratio (λ) for Compression Members:

S. No.	Type of member	λ
1	A strut connected by single rivet at each end	180
2	A member carrying compressive loads resulting from dead loads and imposed loads	180
3	A member subjected to compressive forces resulting from wind / earthquake forces, provided the deformation of such members does not adversely effect the stress in any part of the Structure.	250
4	Compression flange of a beam	300
5	A member normally acting as a tie in a roof truss or a bracing system but subjected to possible reversal of stresses resulting from the action of wind or earthquake forces.	350

2.2.3 Types of Cross-Section

Requirements for compression members are more demanding than those for tension members, for here the carrying capacity is a function of the shape as well as the area and also the material properties. The material must be disposed so as to resist effectively any tendency towards general or local instability. Thus the member must be sufficiently rigid to prevent general buckling in any possible direction and each plate element of the member must be thick enough to prevent local buckling.

The most important property of the section in a compression member is the radius of gyration and thus the moment of inertia can be increased by spreading the material of section away from its axis. An ideal section is one which has the same moment of inertia about any axis through its centre of gravity.

Rolled steel sections cost less than the built-up sections per unit weight and are therefore, preferred. Rods and bars withstand very little compression when length is more. Hence these are recommended for lengths less than 3 m only.

Tubular sections are most suitable for small loads and lengths. These sections are usually provided for roof trusses and bracings. The use of tubes as compression members was limited over decades due to the difficulty in making connections with rivets/bolts.

But with the development of welding techniques its use has become frequent for the following reasons:

- Tubes have the same radius of gyration in all directions and have a high local buckling strength.

- Tubes have more torsional resistance.

- In case of members subjected to wind, round tubes are subjected to less force than flat sections.

The least radius of gyration of a single angle section is small as compared to channels and I-Sections and hence, it is not suitable for long lengths. These are therefore, more commonly used as strut in roof trusses with riveted or welded connections. Single angle sections in general should therefore be avoided, wherever possible. Equal angles are more desirable and economical than unequal ones, because their least radius of gyration is greater for the same area of steel.

Where compression members are designed for very large structures it may be desirable to use built-up sections. When these are used they must be connected suitably on their open sides to hold the parts together in their proper positions and to assist them in acting together as an integral unit.

Double angles placed back to back (figure (a)) or with legs spread (figure (b)) are most suited for trusses. For a double angle section back-to-back, it is desirable to use unequal angles with the long legs back-to-back to achieve a balance between radius of gyration values about the X and Y axis.

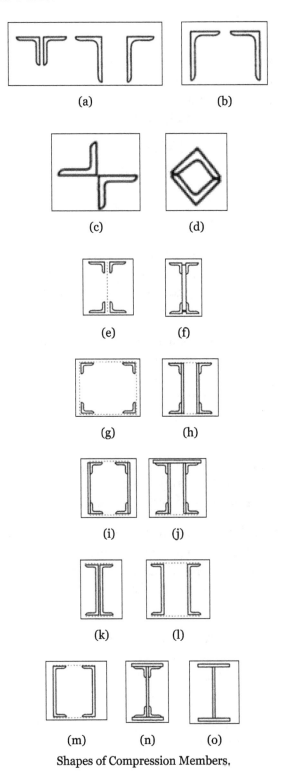

Shapes of Compression Members.

When the legs are spread, the least radius of gyration is same as that when placed back-to-back. The stiffness of double angle sections with legs spread is greater than when placed back-to-back but this is uneconomical because of the additional cost of lacing and riveting.

However, it has no advantage over a double angle placed back-to-back regarding the least radius of gyration. Two angle sections can also be used in the form of a star (figure(c)). This cruciform arrangement of double angles is most effective because of its approximately equal radii of gyration in two directions.

The box section can be formed with welded connection (figure (d)). The least radius of gyration of the section shown in figure(c) and (d) of two equal angles is same and is more than that of the sections shown in figure (a, b). Four angle sections can be used as shown in figure (e-g). The built-up section shown in figure (e) is provided when the load is small. For higher loads the section shown in figure (f) is preferred.

In order to obtain higher values of least radius of gyration in shapes (e) and (f), the longer legs of unequal angles should be kept horizontal, as indicated. The section shown in figure (g) is used for moderate loads and large length, i.e. when the least radius of gyration required is large but due to excessive lacing it is uneconomical.

The connecting systems can be replaced by plates on one side or on more than one side, when a greater cross-sectional area is required or where stiffer bracing between the parts is necessary, as shown in figure (h, i, j). Such sections are normally provided in bridge trusses. Two channels placed back-to-back (figure (l)) result in a small value of radius of gyration about y-y axis and are, therefore, seldom recommended.

Two channels spaced apart (figure (I)) constitute a good section. Two channels placed face to face (figure (m)) provide a larger value of radius of gyration as compared to channels back-to-back (figure (l)) and separated apart for the same spacing.

Thus two channels face to face are the ideal section for compression members as these provide more rigidity. However, when a large radius of gyration is not the criteria for the type required, channels are placed back-to-back as the lacing is minimized. Also the exposed location of both ends of each rivet makes it easier to fabricate and further lower the cost.

Channels placed face to face are also called placed toe-to-toe or placed with flanges turned inwards. The dotted lines in the sections shown in figure represent connecting systems such as lacings, etc. or discontinuous parts and the solid lines represent parts that are continuous for the full length of members.

Generally rolled I.S.H.B. section with additional intermediate support in the weak direction are used as column sections. For large loads, when rolled I-Section do not suffice, an I-Section can be built up with plates and angles in case riveted connections are done (figure(f) and (n)). On the other hand, such a section is built up with plates only if welded connections are done (figure(o)). Two I-Sections laced together can also be used for higher loads.

2.2.4 Classification of Cross-Section

In order to assess the extent to which the resistance and rotation capacity of a cross section is limited by its susceptibility to local buckling without calculating the local buckling resistance, the cross sections for compression member are classified as:

- Class 1 (Plastic compression member): A cross sections are those which can form a plastic hinge and have the (inelastic) rotation capacity required for failure of the structure by formation of plastic mechanism. In other words, the cross sections have sufficient plastic hinge rotation capacity to allow redistribution of moments within the structure.

- Class 2 (Compact compression member): A cross sections are those which are capable of developing a fully plastic stress distribution, i.e., the plastic moment capacity, but local buckling may prevent the development of plastic hinge with sufficient rotation capacity at the section. In other words, these sections have limited plastic hinge (inelastic) rotation capacity for formation of plastic mechanism.

Plastic means that the section is stressed throughout to the yield stress. For a section to be compact, its compression elements shall be continuously connected to the web and the width to thickness ratio of plate elements shall be equal to less than the limiting values specified under Class 2 (Compact), but greater than that specified under Class 1 (Plastic) in table:

- Class 3 (Semi-compact compression member): These are the cross sections in which the stress in the extreme compression fibre of section (assuming elastic distribution of stresses) can reach the yield strength, but the section is not capable of developing the plastic moment of resistance due to local buckling effects.

The width to thickness ratio of plate elements should be equal to or less than that specified under Class 3 (Semi-compact), but greater than that specified under Class 2 (Compact) as given in table:

- Class 4 (Slender compression member): These are the cross sections in which one or more parts of the cross section buckle locally even before reaching yield stress. The reduction in design stress is severe. In such cases, to account for the post local buckling strength of the cross section, the width of the compression plate element in excess of the semi compact section limit is deducted.

As a result, it is usually more economical to thicken the members to take them out of slender range. The width to thickness ratio of plate elements shall be greater than that specified under Class 3 (Semi-compact), as given in table.

When different compression elements of a cross section fall under different classes, the section should be classified as governed by the highest (least favourable) class of its compression elements, i.e., the most critical element. Alternatively, a cross section may be classified with its compression flange and its web in different classes.

The maximum value of limiting width-to-thickness ratios of elements for different classifications of sections are given in Table of the Code.

Table: Limiting width-to-thickness ratio values as recommended in the IS: 800:

Compression element of cross section			Ratio	Class of section		
				Class 1 Plastic	Class 2 Compact	Class 3 Semi-compact
Outstanding element of compression flange.	(a) Rolled steel sections.		b/t_f	9.4ε	10.5ε	15.7ε
	(b) Welded sections.		b/t_f	8.4ε	9.4ε	13.6ε
	Web of a channel.		h/t_w	42ε	42ε	42ε
	Angle subjected to compression due to bending (Both criteria should be satisfied).		b/t d/t	9.4ε	10.5ε	15.7ε
	Single angle or double angles with the components separated subjected to axial compression (All three criteria should be satisfied).		b/t d/t $(b+d)/t$	Not applicable		15.7ε 15.7ε 25.0ε
	Outstanding leg of an angle in contact back to back in a double angle section of a member.		d/t	9.4ε	10.5ε	15.7ε
	Outstanding leg of an angle with its back in continuous contact with another component.		d/t	9.4ε	10.5ε	15.7ε
	Stem of a tee section, rolled or cut from a rolled I-or H-section.		h/t	8.4ε	9.4ε	18.9ε
	Tubular hollow tube (including welded tube).	(a) Moment	D/t	$42.0\varepsilon^2$	$52.0\varepsilon^2$	$88.0\varepsilon^2$
		(b) Axial compression	D/t	Not applicable		$88.0\varepsilon^2$

Where,

ε = Yield stress ratio = $\sqrt{(250/f_y)}$

2.3   Design of Axially Loaded Compression Members

Following are the assumptions made while designing a column:

- The ideal column is assumed to be absolutely straight having no crookedness, which never occurs in practice.

- The modulus of elasticity is assumed to be constant in a built-up column.

- Secondary stresses (which may be of the order of even 25% to 40% of primary stresses) are neglected.

The cross-sectional shape of an axially loaded column depends largely on the length and load on the column. In design information length of the member will be required the end conditions and the loads it has to support. The designer is supposed to select a section which provides a large radius of gyration without providing more area and in which the average compressive stress does not exceed the allowable compressive stress.

To compute these stresses, the cross-sectional area and the radius of gyration must be known. So there are two unknowns and it becomes essential to assume one out of the two based on some principles and compute the other. The section is then checked for safety.

The procedure is thus of trial and error and is as follows:

- Average allowable compressive stress in the section is assumed. It should not be more than the upper limit for the column formula specified by the relevant code.

- The cross-sectional area required to carry the load at the assumed allowable stress is computed.

    A= P / Allowable compressive stress

Where,

A - Tentative cross-sectional area required (in mm²)

P - Load on column in Newtons.

- A section that provides the estimated required area is selected from I.S. The section is so chosen that the minimum radius of gyration of the section selected is as large as possible. The appropriate least radius of gyration for the section selected is recorded.

- The effective length of the column is calculated on the basis of end conditions, and the slenderness ratio is computed ($\lambda = l / r$), which should be less than the permissible slenderness ratio.

- For this estimated value of slenderness ratio, the maximum allowable compressive stress, σ_{ac} is calculated. In case of struts this value may have to be reduced to 80 per cent.

- The load carrying capacity of the member is computed by multiplying the maximum compressive stress thus obtained with the cross sectional area provided. This value of the load carrying capacity of the member should be more than the load to be supported by it.

Column Design Methodology

- Make assumptions about the limiting stress from: (i) buckling stress (ii) axial stress (iii) combined stress.

- Find values for r or A or S ($=I/c_{max}$).

- Pick a trial section based on if we think r or A is going to govern the section size.

- Analyze the stresses and compare using the allowable stress method or interaction formula for eccentric columns.

- The section should pass the stress test.

- Change the section choice and go back to step 4.

Repeat until the section meets the stress criteria to design a simple compression member it is first necessary to evaluate its two effective lengths in relation to the two principal axis, bearing in mind the expected connections at its end.

The verification procedure should then proceed as follows:

- Geometric characteristics of the shape and its yield strength give the reference slenderness λ.

- χ - is calculated taking into account the forming process and the shape thickness using one of the buckling curves and λ.

The design buckling resistance of a compression member is then taken as,

$$N_{b.Rd} = \chi\beta_A \frac{Af_y}{\gamma_{M1}}$$

Where,

$\beta_A = 1$ for class 1,2,3 cross-sections and $\beta_A = A A_{eff}$

If this is higher than the design axial load the column is acceptable if not, then another larger cross-section must be chosen and checked.

2.3.1 Lacing and Battening

Design of Laced Columns

- As far as possible the latticing system shall be uniform throughout.

- In single laced system the direction of lattices or opposite faces should be shadow of the other. It should not be mutually opposite.

- In bolted/riveted construction, the minimum width of lacing bars shall be three times the nominal diameter of the bolts/rivets.

- The thickness of flat lacing bars shall not be less than $1/40^{th}$ of its effective length for single lacing and $1/16^{th}$ of the effective length for double lacings.

- Lacing bars shall be inclined at 40° to 70° to the axis of built up member.

- The distance b/w the two main members should be kept so as to get $r_{yy} > r_{zz}$

Where,

r_{yy} – R.o.G about weaker axis.

r_{zz} – R.o.G of stronger axis of individual member.

- Maximum spacing of lacing bars shall be such that the maximum slenderness of the main member between consecutive lacing connection is not greater than 50 or 0.7 times the most unfavorable slenderness ratio of the member as a whole.

- The lacing shall be designed to resist transverse shear $v_t = 2.5\%$ of axial force in column. If there are two transverse parallel system then each system has to resist $v_t /2$ shear force.

- If the column is subjected to bending also, $v_t =$ Bending shear + 2.5% column force.

- Effective length of single laced system is equal to the length between the inner end fastener. For welded joints and double laced effectively connected t intersection effective length may be taken as 0.7 times the actual length.

Design of Batten Columns

- Batten plates should be provided symmetrically.

- At both ends batten plates should be provided. They should be provided at point where the member is stayed in its length.

- The number of battens should be such that the member is divided into not less than three bays as far as they should be spaced and proportioned uniformly throughout.

- Battens shall be of plates, angles, channels or I-sections and at their ends shall be riveted, bolted or welded.

- By providing battens distance between the member of columns is so maintained that r yy >r xx

- The effective slenderness ratio of battened columns shall be taken as 1.1 times the maximum actual slenderness ratio of the column to account for shear deformation.

- The vertical spacing of battens, measured as centre to centre of its end fastening shall be such that the slenderness ratio of any component of column over that distance shall be neither greater than 50nor greater than 0.7 times the slenderness ratio of the member as a whole above its z-z axis.

- Battens shall be designed to carry the bending moments and shear force arising from transverse shear force v_t equal to 2.5% of the total axial load.

- In case columns are subjected to moments also the resulting shear force should be found and then the design shear 2.5% of axial load.

- The design shear end moments for battens plates is given by $V_b = V_t$ x C/NS and $M = V_t$ x C/2N at each connection.

Where,

V_t - Transverse shear force.

C - Distance b/w Centre to Centre of battens longitudinally.

N- Number of parallel planes.

S - Minimum transverse distance between the centroid of the fasteners connecting batten to the main member.

- The effective depth of end battens (longitudinally), shall not be less than the distance between the centroids of main members.

- Effective depth of intermediate battens shall not be less than $3/4^{th}$ of the above distance.

- In no case the width of battens shall be less than twice the width of one member in the plane of the batten. It is to be noted that the effective depth of the batten shall be taken as the longitudinal distance between the outermost fasteners.

- The thickness of battens shall be not less than $1/50^{th}$ of the distance between the innermost connecting lines of rivets, bolts or welds.

- The length of the weld connecting batten plate to the member shall not be less than half the depth of batten plate. At least one third of the weld shall be placed at each end of this edge.

Problems

1. Let us design a laced column with two channels back to back of length 10 m to carry an axial factored load of 1400 kN. The column may be assumed to have restrained in position but not in direction at both ends (hinged ends).

Solution:

Given:

Length = 10 m

Load = 1400 kN

Assuming f_{cd} = 135 N/mm²

Area required = $\dfrac{1400 \times 1000}{135} = 10370 \, mm^2$

Try 2 ISMC 350 at 413 N/m.

Area provided = $2 \times 5366 = 10732 \, mm^2$

Distance will be maintained so as to get $r_{yy} > r_{zz}$.

$\therefore$ Actual = $\dfrac{KL}{r} = \dfrac{1 \times 10000}{136.6} = 73.206$

Since it is a laced column,

$\dfrac{KL}{r} = 1.05 \times 73.206 = 76.87$

From Table,

$$f_{cd} = 152 - \frac{6.87}{10}(152 - 136)$$
$$= 141.0 \text{ N/mm}^2$$

Load carrying capacity,

$$= 10732 \times 141.0$$

$$= 1513.297 \times 10^3$$

$$= 1513.297 \text{ kN} = 1400 \text{ kN}$$

Hence, design is safe.

Spacing between the channels.

Let it be at a clear distance "d".

Now,

$$I_{xx} = 2 \times 10008 \times 10^4 = 20016 \times 10^4 \text{ mm}^4$$
$$I_{yy} = 2\left[430.6 \times 10^4 + 5366\left(\frac{d}{2} + 24.4\right)^2\right]$$

Equating I_{yy} to I_{xx}, we get,

$$2\left[430.6 \times 10^4 + 5366\left(\frac{d}{2} + 24.4\right)^2\right] = 20016 \times 10^4$$

i.e.,

$$\left(\frac{d}{2} + 24.4\right)^2 = 17848.3$$

Provide d = 220 mm as shown in figure.

Let the lacings be provided at 45° to horizontal.

Horizontal spacing of lacing,

$$= 220 + 60 + 60$$

$$= 340 \text{ mm [Note: g = 60 is gauge distance].}$$

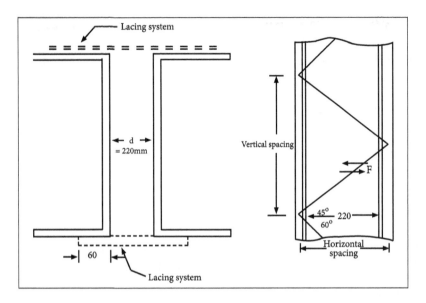

Least r of one channel = r_{yy} = 28.3

Transverse shear to be resisted by lacing systems,

$$= \frac{2.5}{100} \times 1400 \times 10^3 = 35000\,\text{N}$$

Shear to be resisted by each lacing systems,

$$= \frac{35000}{2} = 17500\,\text{N}$$

Length of lacing $= (220 + 60 + 60)\dfrac{1}{\cos 45} = 480.83$ mm

Minimum thickness of lacing,

$$= \frac{1}{40} \times 480.83$$

$$= 12.02 \text{ mm}$$

Sectional area = 60 × 14 = 840 mm²,

$$r_{min} = \sqrt{\frac{\frac{1}{12} \times 60 \times 14^3}{60 \times 14}} = 4.041 \text{ mm}$$

$$\frac{KL}{r} = \frac{480.3}{4.641} = 18.97 \times 14.5$$

Strength of 20 mm Shop Bolts

(i) In single shear:

$$= 0.78 \times \frac{\pi}{4} \times 20^2 \times \frac{400}{\sqrt{3} \times 1.25} = 45272\,N$$

$$\text{Edge distance} = \frac{60}{2} = 30$$

$$\therefore K_b = \frac{30}{3 \times 22} = 0.4545$$

(ii) Strength in bearing:

$$= \frac{2.5\,K_b\,dt\,f_u}{1.25}$$

$$= \frac{2.5 \times 0.4545 \times 20 \times 10 \times 400}{1.25}$$

$$= 101808\,N$$

Bolt value = 45272

$$\text{Number of bolt required} = \frac{17500}{45272} = 0.387$$

Provide one bolt.

Check for the strength of lacing,

$$\frac{KL}{r} = 118.97$$

A flat belongs to bucking class c.

$$f_{cd} = 94.6 - \frac{8.97}{10}(94.6 - 83.7)$$
$$= 84.82\,N/mm^2$$

Load carrying capacity in compression,

$$= 84.82 \times 60 \times 14$$

$$= 71251\,N$$

2. Let us design a laced column for an axial load of 1200 kN with an effective span of 7.5

m has one end fixed out the other end hinged. Use channels for main members and an angle for lacing bars.

Solution:

Given:

> Axial load =1200 kN

> Span =7.5 m

Step 1: Let us assume f_{cd} = 135 MPa

Step 2: Gross area required for the column $\dfrac{P_d}{f_{cd}}$

Design load for the column,

> = (1.5) (1200)

> = 1800 kN

Gross area required,

$$= \frac{1800 \times 1000}{135}$$

> = 13333.33 mm²

Step 3: Let us assume that the channels are placed back to back.

Let us try with 2 channels 23 MC 400 at 484.5 N/N

Giving an area of 2(6293) = 12586 mm²

γ_{xx} of one 1 SMC 400 = 154.8 mm

Least radius of gyration γ_{xx} = 154.8 mm

The effective length KL is given as = 7500 mm

$$\frac{KL}{\gamma} = \frac{7500}{*} = *$$

Design the two channels as connected by Larcey's,

$$\left(\frac{KL}{\gamma}\right)_e = 1.05\left(\frac{KL}{\gamma}\right)_o$$

$$\left(\frac{KL}{\gamma}\right)_e = 1.05(41.34)$$

$$= 43.407$$

Step 4: Refer table 10 of IS 800.

$$\frac{KL}{\gamma} = 43.407$$

$$f_y = 250 \text{ MPa}$$

$$f_{cd} = 198 - \left(\frac{198 - 183}{10}\right)3.407$$

$$= 192.89 \text{ MPa}$$

Load carrying capacity $= f_{cd}(A)$

$$= 192.89 \, (12586)$$

$$= 2427714 \text{ N}$$

$$= 2427.714 \text{ kN}$$

Step 5: Spacing between the channels.

From steel table for one 1 SMC,

$$I_{XX} = 15082.8 \times 10^4 \text{ mm}^2$$

For two channels,

$$I_{xx} = 2 \,(15082.8 \times 10^4) \text{ mm}^4$$

$$I_{YY} = 2\left[504 \times 10^4 + 6293\left(\frac{d}{2} + 29.2\right)\right]$$

$$I_{xx} = I_{xx}$$

$$d = 256 \text{ mm}$$

Let us provide 260 mm as the distance between 2 channels.

Step 6: Design of lacings.

Let the lacings be provided at 45° to the horizontal and let the gauge distance for the bolt be 60 mm.

Horizontal spacing of lacings = 260 + 60 + 60

$$= 380 \text{ mm}$$

The vertical spacing (1) = 2 (380 t an 45°)

    = 760 mm

Least value of γ_{min} = 28.3 mm for ISMC 4 N,

$$\frac{a_1}{n} > 50$$

$$\frac{a_1}{n} = \frac{760}{28.3} = 26.855 < 50$$

Transverse shear to be resisted by lacing systems = 2.5%,

Axial force in the column,

$$= \frac{2.5}{100}\left(1800 \times 10^3\right)$$

    = 45000 N

Shear to be resisted by each lacing,

$$= \frac{45000}{2} = 22500\,N$$

Length of Lacing,

$$= (260 + 60 + 60)\,\frac{1}{\cos 45°}$$

    = 537.40 mm

Minimum width of lacing as per IS 800- 2007.

    = 3 × (nominal diameter of end bolt)

    = 3 × 20

    = 60 mm (assuming 20 mm ϕ bolt)

Minimum thickness of lacing as per IS 800 -2007.

$$= \frac{1}{40}\,\text{(Effective length of lacing)}$$

$$= \frac{1}{40}(537.40)$$

    = 13.435 mm

Let us provide 14 mm thick flats for the lacing.

Hence let us use 60 ISF 14 for lacing.

Sectional area = 60 × 14

$$= 840 \text{ mm}^2$$

$$\gamma_{min} = \sqrt{1/A}$$

$$= \sqrt{\frac{60 \times 14^3/12}{60 \times 14}} = 4.041 \text{ mm}$$

$$\frac{KL}{\gamma} = \frac{L}{r} = \frac{537.40}{4.041} = 132.99 < 145$$

Hence, design is safe.

Step 7: Design of bolt for the lacings.

Assuming 20 mm bolts

Strength of one bolt in single shear.

$$= 0.78 \left(\frac{\pi (20)^2}{4} \right) \frac{400\sqrt{3}}{1.25} = 45272 \text{ mm}$$

Edge distance,

$$= \frac{60}{2} = 30 \text{ mm}$$

$$k_b = \frac{P}{3d_o} = \frac{30}{3(22)} = 0.4545$$

Strength of one bolt in bearing.

$$= \frac{2.5 \, k_b \, dt \, f_u}{1.25}$$

$$= \frac{2.5(0.4545)(20)(14)(410)}{1.25} = 104353 \text{ N}$$

Bolt Value = 45272 N

$$\text{No. of bolts} = \frac{18750}{45272} = 0.414$$

Hence let us provide one bolt for the connection at each end of the lacing.

Step 8: Check for strength of lacing,

$$\frac{KL}{\gamma} = 132.99$$

Referring to Table 10 of IS 800 : 2007.

$$f_{cd} = 74.3 - \frac{74.3 - 66.2}{10}$$

$$= 2.99$$

$$f_{cd} = 71.88 \text{ MPa}$$

Load carrying capacity in compression of lacing,

$$f_{cd} = 71.88 \, (60 \times 14)$$

$$= 60379 \text{ N}$$

3. Let us design a column with single lacing system that carry a factored axial load of 1500 kN. The effective net of the column is 4.2 m. Use two channels placed toe to toe.

Solution:

Given:

    Load = 1500 kN

    Column = 4.2 m

Step 1: Let us assume the value of f_{cd} = 135 MPa

Step 2: Gross area assumed for the column $A_g = \dfrac{P_d}{f_{cd}}$

    Axial load P_d = 1500 kN

    $\therefore$ Gross area required $= \dfrac{1500(1000)}{135} = 1111 \text{ mm}^2.$

Step 3: γ_{xx} of one IS MC = 154.8 mm

Least radius of gyration = γ_{xx} = 154.8 mm

$\therefore$ The effective length,

$$\frac{KL}{\gamma} = \frac{4200}{154.8} = 27.13$$

IS 800 – 2007,

$$\left(\frac{K_L}{\gamma}\right)_2 = 1.05\left(\frac{K_L}{\gamma}\right)_0$$

Where,

$$\left(\frac{K_L}{\gamma}\right)_e = 1.05(27.13) = 28.487$$

Step 4: $f_{cd} = 224 - \dfrac{(224-211)}{10}(8.487)$

$\qquad = 212.97 \text{ MPa}$

Load carrying capacity = $f_{cd}(A)$

$$= 212.97\,(12586)$$

$$= 2680440 \text{ N}$$

$$= 2680.44 \text{ kN} > 1500 \text{ kN}$$

Step 5: $I_{XX} = (15082.8 \times 10^4)$ mm^4, $C_{XX} = 24.2$ mm

$\qquad I_{XX} = 2(15082.8 \times 10^4)$ mm^4

$$I_{yy} = 2\left[504.8\times10^4 + 6293\left(\frac{d}{2}-24.2\right)^2\right] = 2\left(15082.8 \times 10^4\right)$$

$\qquad d = 352.8$ mm

Step 6: Design of lacings

$\therefore$ Horizontal spacing of lacings = 360 + 60 + 60 = 480 mm

$\therefore$ The vertical spacing of lacing bar,

$$= 2(360 - 60 - 60)\,(\tan 45°)$$

$$= 2(240 \tan 45°) = 480 \text{ mm}$$

$$\frac{a_1}{r_1} = \frac{480}{28.3} = 16.96 < 50$$

Transfer shear system = 2.5%,

Axial force column,

$$= \frac{2.5}{100}\left(1500\times10^3\right) = 37500\text{N}$$

$\therefore$ Shear resisted $= \dfrac{37500}{2} = 18750\,N$.

Minimum Width of Lacing:

$3 \times$ (nominal diameter) $= 3 \times 20 = 60$ mm

Minimum Thickness:

$$= \dfrac{1}{40} \text{ (effect length)}$$

$$= (360 - 50 - 50) = \dfrac{1}{\cos 45°}$$

$$= 367.7 \text{ mm}$$

$$\gamma_{min} = \sqrt{I/A}$$

$$= \sqrt{\dfrac{60 \times 10^3/12}{60 \times 10}} = \dfrac{t}{\sqrt{12}} = \dfrac{10}{\sqrt{12}}$$

$$\dfrac{K_L}{\gamma} = \dfrac{l}{r} = \dfrac{367.7}{2.89} = 127.33 < 145$$

Step 7: Design of bolt for the lacing

$$= 0.78 \left(\dfrac{\pi (20)^2}{4} \right) \dfrac{400\sqrt{3}}{1.25}$$

$$= 45272 \text{ N}$$

$$= \dfrac{60}{2} = 30 \text{ mm}$$

$$k_b = \dfrac{P}{3d_o} = \dfrac{3}{3(22)} = 0.4545$$

Strength of one bolt bearing,

$$= \dfrac{2.5\,k_b\,dt\,fy}{1.25}$$

$$= \dfrac{2.5(0.4545)(20)(10)(410)}{1.25}$$

$$= 45272 \text{ N}$$

No. of bolt $= \dfrac{18750}{45272} = 0.414$

$$\frac{k_L}{\gamma} = 127.23$$

Check for strength of lacing,

$$f_{cd} = \frac{88.7(83.7 - 7413)(-1.23)}{10} = 76.90 \text{ Mp}$$

Load carrying capacity = f_{cd} (area of cross section)

$$= 76.90 \ (60 \times 10) = 46142 \text{ N}$$

Force in lacing $= \dfrac{18750}{\sin 45°} = 26516 \text{ N} < 46142 \text{ N}$

∴ Let us provide 60 IS flat Inclination and connect them to centre of gravity of channels with one bolt of 20 mm nominal diameter at each end.

2.3.2 Design of Column Bases and Foundation Bolts

Steps in the Design of a Column Slab Base Connection:

- Maximum bearing pressure of concrete for the slab base 0.6 fck.
- Area of base plate.
- Thickness of base plate.
- Connection.

The commonly used types of column bases are:

- Slab base
- Gusseted base.

For columns with gusseted bases, all the bearing surfaces are machined or milled to ensure perfect contact. Where the ends of the column shaft and the gusset plates are not faced for complete bearing, the welding, fastenings connecting them to the base plate should be sufficient to transmit all the forces to which the base is subjected.

Where the end of the column is connected directly to the base plate by means of full penetration butt welds, the connection is designed to transmit to the base all the forces and moments to which the column is subjected. The nominal bearing pressure between the base plate and the supporting concrete block can be determined by assuming a linear distribution.

The maximum stress induced on concrete foundation should not be larger than 0.6 fa in which fa is the lesser of 28-day cube strength of the concrete or the bedding grout.

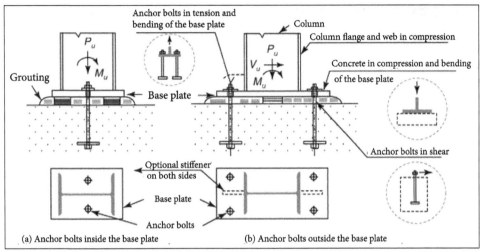

Components for typical column based assembly.

Design of Foundation Bolts

Foundation bolts are provided to resist the uplift on the windward side. The maximum uplift force per unit length of circumference at the base on the windward side,

$$F_t = \frac{4M_w}{\pi d_c^2} - \frac{W_s}{\pi d_c} \, kN/m$$

In the above expression, weight of lining has been deliberately ignored.

Let g be the spacing of the bolts, in meters. Hence force (F_b) per bolt is,

$$F_b = F_t \cdot g = \left(\frac{4M_w}{\pi d_c^2} - \frac{W_s}{\pi d_c} \right) g \, kN$$

The number of bolts may be determined by dividing the circumference of anchor bolt ring (of diameter d_b, say) by the spacing g of the bolts.

Problems

1. A column of ISMB 400 is subjected to an axial force of 750 kN. Let us design suitable base plate and assume necessary data's required.

Solution:

Given:

Axial force of 750 kN

Design:

Step 1: IS 800: 2007,

Bearing strength of concrete pedestal,

$$= 0.45\, f_{cl}$$

$$= 0.45\,(20)$$

$$= 9\, \text{MPa}$$

Assuming a load factor 1.5,

Factored load $p_u = 1.5\,(750)$

$$= 1125\, \text{kN}$$

Step 2: Area of base plate,

Area of base plate required,

$$= \frac{1125 \times 1000}{9} = 125000\, \text{mm}^2$$

The properties of IS MB 400

D = 400 mm

B = 140 mm

t_f = 16 m

t_w = 8.9 m

Provide an equal projection of for all the 4 edges of the column.

Size of base plate provided = [400 + 2(60)] × (140 + 2(f$_o$))

$$\Rightarrow 520\, \text{mm} \times 260\, \text{mm}$$

Area of base provided = 135200 mm²

Step 3: Thickness of base plate,

Actual pressure at base $W = \dfrac{7125 \times 10^3}{520 \times 260}$

W = 8.32 MPa

$$t_s = \sqrt{2.5\,w\left(a^2 - 0.3b^2\right)\gamma_{mo}\, fy} > tf$$

W = 8.32 MPa, a = b = 60 mm, γ_{mw} = 1.1, f_y = 250 MPa

t_f = 16 mm for given ISMB 4 av

$$t_s = \sqrt{2.5(7.32)\left(60^2 - 0.3(60)^2(1.1)/250\right)} = 15.18 > 16\,mm$$

= 15.18 mm should be t_7.

Hence let us provide a thickness of 16 mm for base plates.

Step 4: Connection.

Let us use 4 bolts of 20 mm diameter and 300 mm long to anchor the blade of size.

520 × 260 × 16 mm to the concrete formulation.

Welds: The length available for welding.

= 2 [140 + (140 − 16) + 400 − 2(7.5)] = 1298 mm

Strength of weld/mm = $\dfrac{410\sqrt{3}}{1.25}$ = 189.3 MPa

Effective area of weld = 0.7 (t) (L_e)

0.7 (t) (L_e) (189.38) = 1125 × 10³

$$t = \frac{1125 \times 10^3}{0.7 \times 189.38 \times 1298}$$

t = 6.53 mm

Hence let us provide 8 mm as the size of weld.

2. Let us design a steel column of rolled steel I section which carry an axial load of 500 kN. The column is 4 m long and it is effectively held in position at both ends but restrained against at one end only. Take yield stress in steel as 250 N/mm².

Solution:

Given:

      Axial load of 500 kN

      Long =4 m

      Yield stress in steel =250 N/mm²

Design a Steel Column:

Rolled Steel I Section.

Axial load, P_d = 500 kN.

Column = 4 m long

Effectively held at both ends in position one end is restrained,

$$f_y = 250 \, N/mm^2$$

∴ Selection of Suitable Section:

$$\left.\begin{array}{l}\text{One end restrained} \\ \text{Other end held in position}\end{array}\right\} \Rightarrow \text{KL Effective Length} = 1.2\,L$$

Refer as per Table 11 – IS 800:2007

Effective Length = 1.2L = 1.2 × 4000,

$$= 6000 \, mm$$

Assume design stress for an of section f_{cd} = 150Mpa

$$A_g = A = \frac{P_d}{f_{cd}} = \frac{500000}{150} = 3333 \, mm^2$$

Gross Section Area Required

By referring steel table choosing, ISMB 225

$$A_g = 39723 \, mm^2$$

$$h = 225 \, mm$$

$$b_f = 110 \, mm$$

$$L_c = 11.9 \, mm$$

$$t_w = 6.5 \, mm$$

$$r_{YY} = 23.4 \, mm$$

Buckling Class

As per Table of IS800-2007.

$$\frac{h}{b_f} = \frac{225}{250} = 0.9 \, \& \, t_f = 11.8 \, mm \le 40mm$$

$$\frac{h}{b_f} \le 1.2 \, t_f \le 100 \, mm$$

It falls under buckling Class 'b' for buckling about ZZ-axis and buckling Class 'c' for buckling about YY-axis.

Computation of Euler Buckling Stress

$$f_{cc} = \frac{\pi^2 E}{\left[\dfrac{KL}{r_{min}}\right]^2} = \frac{\pi^2 \times 2 \times 10^5}{\left(\dfrac{4000}{23.4}\right)^2} = 67.55 \, \text{N/mm}^2$$

$$= 67.55 \, \text{N/mm}^2$$

Calculation of Euler Slenderness Ratio, λ

As per Cl.7.1.2.1 of IS800-2007

$$\lambda = \sqrt{\left(\frac{f_y}{f_{cc}}\right)} = \sqrt{\frac{250}{67.5}} = 1.92$$

Computation of f_{cd}:

As per Cl.7.1.2.1 of IS800-2007.

$$f_{cd} = \frac{f_y / \gamma_{m_0}}{\phi + \left[\phi^2 - \lambda^2\right]^{0.5}} \le \frac{f_y}{\gamma_{m_0}}$$

$\phi = 0.5[1 + \alpha(\lambda - 0.2) + \lambda^2]$

$\alpha = 0.34$ [Table of IS800-2007]

$\phi = 0.5[1 + 0.34[1.92 - 0.2] + 1.92^2] = 2.6356$

$f_y = 250$ Mpa

$\gamma_{m_0} = 1.1$ [Table of IS800-2007]

$$f_{cd} = \frac{250/1.1}{2.6356 + \left[2.6356^2 - 1.92^2\right]^{0.5}} \le \frac{250}{1.1}$$

$$= 51.17 \le 227.27 \, \text{MPa.}$$

Computation of Strength of Column

$$P_d = A_g f_{cd}$$

$$= 3972 \times 51.17$$

$$= 203247 \, \text{N}$$

3. Let us design a buildup column composed of two channel sections placed back to back, carrying an axial load of 1345 kN. Effective length of column is 5.95 m. Take f_y = 250 kN/mm².

Solution:

Given:

Design a Buildup Column

Two Channel Section – back to back.

Axial Load = 1345 kN

Effective length of column = 5.95 m.

f_y = 250 kN/mm²

Assume f_{cd} = 135 MPa

Gross Section Area Required

$$A = \frac{}{f_{cd}} = \frac{1345 \times 1000}{135} = 9963\,mm$$

9963 mm

Let us try with 2 channel ISMC 400 at 484.5 N/m

Giving an area of 2[6293] = 12586 mm²

γ_{xx} of one ISMC 400 = 154.8 mm

Spacing between the channels will be maintained such that $\gamma_{yy} \geq Y_{xx}$ so that.

Least rate of gyration γ_{xx} = 154.8 mm

Effective Length = KL = 5950 mm

$$\frac{KL}{\gamma} = \frac{5950}{154.8} = 38.4$$

$$= 38.4$$

Design the two channels, as connected by Lacing.

As per CI.7.6.1.5 of IS800-2007.

$$\left(\frac{KL}{\gamma}\right)_e = 1.05\left[\frac{KL}{\gamma}\right]_o$$

$$= 1.05[38.4] = 40.35$$

Referring to Table of IS800-2007, built up Sections belong to buckling Class 'C'.

$$\frac{KL}{\gamma} = 38.4$$

$$f_y = 250 \text{ Mpa}$$

From Table C of IS800-2007.

$$f_{cd} = 201 \text{ MPa.}$$

Load carrying capacity = f_{cd} × Area = 201 × 12586 = 2427714 N

Spacing between the Channels

Let 'd' be the distance between the channel.

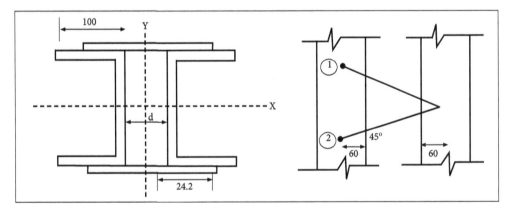

From Steel Table per one ISMC 400,

$$I_{yy} = 15082.8 \times 10^4 \text{ mm}^4$$

$$I_{xx} = 2[504.8 \times 10^4 + 6293 \, [d/2 + 24.2]^2]$$

$$I_{YY} = I_{XX}$$

$$2\left[504.8 \times 10^4 + 6293 \left(\frac{d}{2} \times 24.2\right)^2\right] = 2 \times 15082.8 \times 10^4$$

d = 256 mm ≈ 260 mm

Design of Lacings

Let the Lacings be provided at 45° to the horizontal and let the gauge distance for the bolt be 60 mm.

Horizontal Spacing of Lacing = 260 + 60 + 60

$$= 380 \text{ mm}$$

Vertical Spacing (1), (2) = 2 × 380 × tan 45°

$$= 760 \text{ mm}$$

$$\gamma_{min} = 28.3 \text{ mm}$$

Referring to CI.7.6.5.1,

$$\frac{a_1}{r_1} > 50$$

$$\frac{760}{28.3} > 50$$

$$26.85 > 50$$

As per CI.7 6.6 1 of IS800-2007.

Transverse Shear to be resisted by Lacing System = 2.5% of axial forces in column,

$$= \frac{2.5}{100} \times 1345 \times 10^3 = 33625 \text{ N}$$

∴ Shear to be resisted by each Lacing = 16812.5 N

Length of Lacing = [260 + 60 + 60] $\dfrac{1}{\cos 45°}$ = 537.4 mm

As per CI.7.6.2 of IS 800-2007.

Minimum width of Lacing = 3 × d_n

$$= 3 \times 20 = 60 \text{ mm}$$

[assume 20 mm and bolt]

As per CI.7.6.3 of IS800-2007.

Minimum thickness of lacing = 3 × d_n [assume 20 mm and bolt]

As per CI.7.6.3 of IS800-2007.

Minimum Thickness of Lacing

$$= \frac{1}{40} \text{ Effective length of Lacing}$$

$$= \frac{1}{40} \times 537.4$$

$$= 13.435 \approx 14 \text{ mm}$$

Sectional Area $= 60 \times 14 = 840 \text{ mm}^2$

$$\gamma_{min} = \sqrt{I/A} = \sqrt{\frac{60 \times 14^3}{12 \times 14}} = 4 \text{ mm}$$

By referring to Cl.7.6.6.3 of IS800-2007.

$$KL = L$$

$$\frac{L}{r} < 145$$

$$\frac{537.40}{4.041} = 132.99 < 145$$

Design of Bolt for the Lacings

Assume ϕ 20 mm bolt.

Strength of one bolt in single shear $= 0.78 \left[\frac{\pi}{4} 20^2 \right] \frac{400}{\sqrt{3} \times 1.25} = 45272 \text{N}$

Edge distance $= \frac{60}{2} = 30 \text{ mm}$

$$K_b = \frac{R}{3d_o} = \frac{30}{3 \times 22} = 0.4545$$

Strength of one bolt in bearing,

$$= \frac{2.5 K_b \, dt \, f_u}{1.25}$$

$$= \frac{2.5 \times 0.4545 \times 20 \times 16 \times 410}{1.25}$$

$$= 104353 \text{ N}$$

Bolt Value $= 45272$

No. of bolts required $= \frac{16812.5}{45272} = 0.37 \approx 1$

Hence let us provide one bolt for the connection at each end of the lacing.

Check for Strength of Lacing,

$$\frac{KL}{r} = 132.99$$

Referring to table of IS800-2007, flat section belongs to buckling class 'c'.

Refer Table C of IS800-2007.

$$f_{cd} = 71.8 \, \text{MPa}$$

Load carrying capacity in compression of Lacing,

$$= f_{cd} \times \text{Area}$$

$$= 71.8 \times 60 \times 14$$

$$= 60379 \text{N}$$

Force in Lacing $= \dfrac{16812.5}{\sin 45°} = 23776 \, A < 60379 \, A$

Hence design is Safe.

Let us provide 60 ISF 14 flats at 45° and connect them to centre of gravity of channel with an bolt of 20 mm nominal diameter at each end.

3

Design of Beams and Plate Girders

3.1　Design of Beams and Lateral Stability of Beams

Beams are structural elements subjected to transverse loads in the plane of bending causing BMs and SFs. Symmetrical sections about z-z axis are economical and geometrical properties of such sections are available in SP 16. The compression flange of the beams can be laterally supported (restrained) or laterally unsupported (unrestrained) depending upon whether restraints are provided or not.

The beams are designed for maximum BM and checked for maximum SF, local effects such as vertical buckling and crippling of webs and deflection. Beams can be fabricated to form different types of c/s for the specific requirements of spans and loadings.

A structural member which supports transverse (Perpendicular to the axis of the member) load is called a beam. Beams are subjected to bending moment and shear force. Beams are also known as flexural or bending members. In a beam one of the dimensions is very large compared to the other two dimensions.

Beams might be of the following types:

Singly or Doubly Reinforced Rectangular Beams

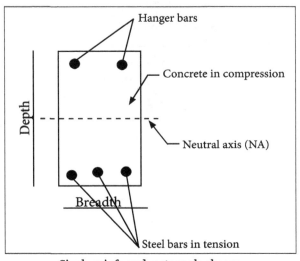

Singly reinforced rectangular beam.

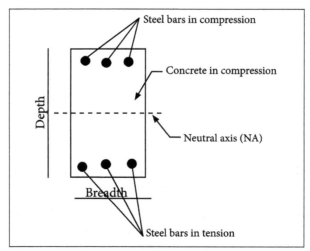

Doubly reinforced rectangular beam.

Singly or Doubly Reinforced T-Beams

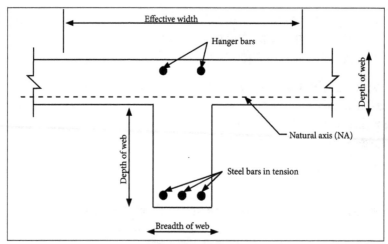

Singly reinforced T beam.

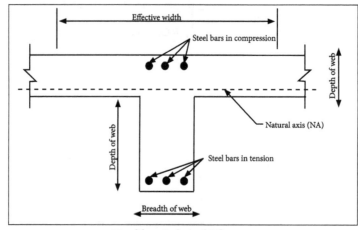

Doubly reinforced T beam.

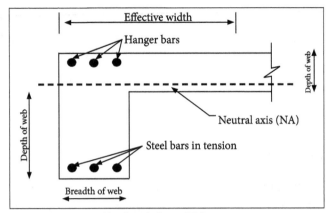

Singly reinforced L beam.

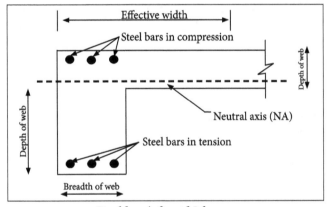

Doubly reinforced L beam.

General Specification for Flexure Design of Beams

Beams are designed on the basis of limit state of collapse in flexure and checked for other limit states of shear, torsion and serviceability. To ensure safety, the resistance to bending, shear, torsion and axial loads at every section must be greater than the appropriate values at that produced by the most unfavorable combination of loads on the structure using the appropriate safety factors.

Selection of grade of concrete apart from strength and deflection, durability shall also be considered to select the grade of concrete to be used. Table of IS 456:2000 shall be referred for the grade of concrete to be used.

Selection of grade of steel normally Fe 250, Fe 415 and Fe 500 are used. In earthquake zones and other places where there are possibilities of vibration, impact, blast etc, Fe 250 (mild steel) is preferred as it is more ductile.

Size of the Beam

The size of the beam shall be fixed based on the architectural requirements, placing of reinforcement, economy of the formwork, deflection, design moments and shear. In addition, the depth of the beam depends on the clear height below the beam and the width depends on the thickness of the wall to be constructed below the beam.

The width of the beam is usually equal to the width of the wall so that there is no projection or offset at the common surface of contact between the beam and the wall. The commonly used widths of the beam are 115mm, 150mm, 200mm, 230mm, 250mm, 300mm.

Cover to the Reinforcement

Cover is the certain thickness of concrete provided all round the steel bars to give adequate protection to steel against fire, corrosion and other harmful elements present in the atmosphere. It is measured as distance from the outer concrete surface to the nearest surface of steel. The amount of cover to be provided depends on the condition of exposure and shall be as given in the Table of IS 456:2000. The cover shall not be less than the diameter of the bar.

Spacing of the Bars

The details of spacing of bars to be provided in beams are given in clause 26.3.2 of IS 456. As per this clause the following shall be considered for spacing of bars.

The horizontal distance between two parallel main bars shall usually be not less than the following:

- Diameter of the bar if the diameters are equal.
- The diameter of the larger bar if the diameters are unequal.
- 5mm more than the nominal maximum size of coarse aggregate.

Greater horizontal spacing than the minimum specified above should be provided wherever possible. However when needle vibrators are used, the horizontal distance between bars of a group may be reduced to two thirds the nominal maximum size of the coarse aggregate, provided that sufficient space is left between groups of bars to enable the vibrator to be immersed.

Where there are 2 or more rows of bars, the bars shall be vertically in line and the minimum vertical distance between the bars shall be of the greatest of the following:

- 15 mm.
- Maximum size of aggregate.
- Maximum size of bars.

Allowable Stresses

The maximum fiber stress in bending for laterally supported beams and girders is $F_b = 0.66 \, F_y$ if they are compact except for hybrid girders and members with yield points exceeding 65 ksi (448.1 MPa). $F_b = 0.60F_y$ for non-compact sections. F_y is the minimum specified yield strength of the steel, ksi (MPa). Table lists values of F_b for two grades of steel.

Yield strength, ksi (MPa)	Compact 0.66 F_y (MPa)	Non-compact, 0.60 F_y (MPa)
36 (248.2)	24 (165.5)	22 (151.7)
50 (344.7)	33 (227.5)	30 (206.8)

The allowable extreme-fiber stress of $0.60F_y$ applies to laterally supported, unsymmetrical members, except channels and to non-compact box sections. Compression on outer surfaces of channels bent about their major axis should not exceed $0.60F_y$.

The allowable stress of $0.66F_y$ for compact members should be reduced to $0.60F_y$ when the compression flange is unsupported for a length, in (mm), exceeding the smaller of,

$$l_{max} = 76.0b_f/\left(F_y\right)^{1/2} \quad l_{max} = 20,000/F_y d/A_f$$

Where,

       b_f - Width of compression flange, in (mm)

       d - Beam depth, in (mm)

       A_f - Area of compression flange, in (mm)²

The allowable stress should be reduced even more when l/r_T exceeds certain limits,

Where, l is the unbraced length, in (mm) of the compression flange and r_T is the radius of gyration in (mm) of a portion of the beam consisting of the compression flange and one-third of the part of the web in compression,

For $\sqrt{102,000C_b/F_y} \le l/r_T \le \sqrt{510,000C_b/F_y}$, use

$$F_b\left[\frac{2}{3} - \frac{F_y\left(l/r_T\right)^2}{1,53,000C_b}\right]F_y$$

For $l/r_T > \sqrt{510,000C_b/F_y}$, use

$$F_b = \frac{170,000C_b}{\left(l/r_T\right)^2}$$

Where,

C_b = Modifier for moment gradient. When however the compression flange is solid and nearly rectangular in cross section and its area is not less than that of the tension flange the allowable stress may be taken as $F_b - 12,000C_b/l_d/A_f$.

When equation applies (except for channels), F_b should be taken as the larger of the values computed from equation above but not more than $0.60F_y$. The moment-gradient factor C_b in equation above may be computed from $C_b = 1.75 + 1.05\ M_1/M_2 + 0.3(M_1/M_2)^2$ less than equal to 2.3. Where M_1 = smaller beam end moment and M_2 = larger beam end moment.

The algebraic sign of M_1/M_2 is positive for double curvature bending and negative for single-curvature bending. When the bending moment at any point within an unbraced length is larger than that at both ends, the value of C_b should be taken as unity.

For braced frames C_b should be taken as unity for computation of F_{bx} and F_{by}. Equations can be simplified by introducing a new term: $Q = (1/r_T)2F_y$ Now, for 0.2 less than equal to Q ess than equal to 1, $F_b = (2 - Q)F_y/3$ For $Q > 1$: $F_b = F_y/3Q$.

Design of Simple and Compound Beams

The design of beams is dependent upon the following factors:

- Magnitude and type of loading.

- Duration of loading.

- Clear span.

- Material of the beam.

- Shape of the beam cross section.

Beams are designed by use of the following formulae:

1. Bending Stress

$$f_w \quad f \quad M_{max}/z$$

Where,

f_w - Allowable bending stress

f - Actual bending stress

M_{max} - Maximum bending moment

z - Section modulus

This relationship derives from simpled beam theory and,

$$M_{max}/I_{NA} = f_{max}/y_{max}$$

And,

$$I_{NA}/y_{max} = Z$$

The maximum bending stress will be found in the section of the beam where the maximum bending moment occurs. The maximum moment can be obtained from the B.M. diagram.

2. Shear Stress

For rectangular cross-sections,

$$t_w^3 \ t = (3 \times Q_{max})/(3 \times A) = 3Q_{max}/2bd$$

For circular cross-sections,

$$t_w^3 \ t = (4 \times Q_{max})/(3 \times A) = 16Q_{max}/3p \ d^2$$

For l-shaped cross-sections of steel beams,

$$t_w^3 \ t \ at \ Q_{max}/A$$

t_w = Allowable shear stress.

T = Actual shear stress.

Q_{max} = Maximum shear force.

A = Cross-section area.

Allowable shear stress like the allowable bending stress differs for different materials and can be obtained from a building code. Maximum shear force is obtained from the shear force diagram.

3. Deflection

In addition, limitations sometimes are placed on maximum deflection of the beam,

$$d_{max} = K_c \times \left(WL^3/EI \right)$$

Design of Composite Beams

The design involves the subsequent aspects:

- Moment capability: The section specified the instant capacity is bigger than that needed. Shear capability to make sure adequate capability has supported the steel section alone as per usual steel style.

- Shear instrumentality capacity: To change full composite action to be achieved. These should be designed to be adequate.

- Longitudinal shear capacity: Check to forestall attainable rending of the concrete on the length of the beam.

Serviceability checks:

- Deflection.

- Elastic behaviour.

- Vibration.

These checks are to make sure the safe and comfy use of the beam in commission. We tend to make sure it doesn't cause cracking of ceilings and isn't dynamically 'lively'. Additionally we tend to verify that it's perpetually elastic once subjected to service masses to avoid issues with plastic strain (i.e. permanent deflection) of the beam. We'll not think about checks on vibration and can solely define the calculations for the elastic check.

Curtailment of Flange Plates

The bending moment diagram for a simply supported beam carrying a UDL varies parabolically throughout the length of the beam. It would then be ideal to vary making the flanges thicker towards the centre of the beam. But for practical reasons this is not possible. In plated beams the flange plates can be curtailed near the supports of the beam resulting in some steel saving.

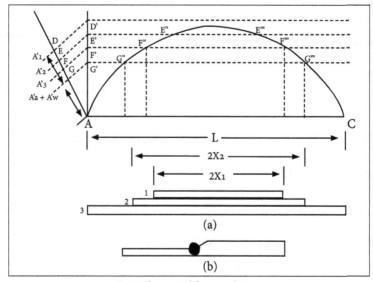

Curtailment of flange plates.

The appropriate points usually termed theoretical cut off points may be found in two ways:

- By algebraic method.

- By geometric construction.

Algebraic Method

Let A'_f net area of flange (flange angle + flange plates) A'_1, A'_2, A'_n = Net areas of 1st, 2nd, n^{th} cover plates respectively counted from outside as the effective depth of the plate girder is approximately same even after curtailment of flange plates it is therefore assumed to be same throughout the span. If the girder is subjected to a uniform load over the entire span the bending moment diagram is parabolic.

$$\frac{wl^2}{8} = \sigma_{bt}\left(A'_1 + A'_{w1}\right)d_1$$

And,

$$\frac{wl^2}{8} - \frac{w\left(2X_n\right)^2}{8} = \sigma_{bt}\left[\left(A'_f + A'_{w1}\right) - \left(A'_1 + A'_2 + ... + A'_n\right)\right]d_1$$

Dividing Equation,

$$1 - \frac{\left(2X_n\right)^2}{l^2} = 1 - \frac{A'_1 + A'_2 + ... + A'_n}{A'_f + A'_{w1}}$$

$$X_n = \frac{1}{2}\sqrt{\frac{A'_1 + A'_2 + ... + A'_n}{A'_f + A'_{w1}}}$$

$$X_1 = \frac{1}{2}\left(\frac{A'_1}{A'_f + A'_{w1}}\right)^{1/2}$$

$$X_2 = \frac{1}{2}\left(\frac{A'_1 + A'_2}{A'_f + A'_{w1}}\right)^{1/2}$$

And x_1, x_2, x_3 are the distances of the cut off points from the centre of the plate girder.

Beam to Beam Connection

The principles of beam-beam connections are the same as those for beam-to-column connection, though some further preparation is also necessary if beams are to be connected with the highest flanges level as is traditional.

Beam-beam connections are usually elaborated so the highest flanges of the two beams are level. The tip of the secondary (supported) beam is so typically notched to modify attachment to thee most (supporting) beam.

The simplest construct for a beam-beam affiliation is to support one beam directly on the highest rim of the opposite. It is often used with the parallel beam system .For different framing arrangements for floors this is often not a standard detail since it ends up in deep construction zones. It's but ordinarily utilized in roof construction wherever purlins acting as secondary structural beams are supported directly on the most beam or rafter.

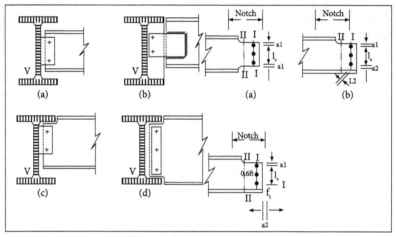

Beam to beam connection.

For I-I:

A net = $t(L_v + l_1 + l_2 - nd)$

n= Number of bolt holes (according to Euro code 3)

$l_1 = 5d$ but $l_1 \leq a_1$

$l_2 = 5d$ but $l_2 \leq a_2$

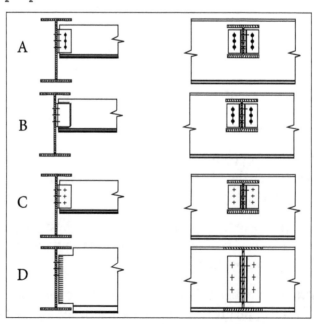

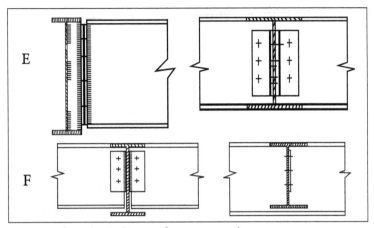

Beam to beam connection.

Types of Beam Cross Sections

The moment resisting capacity of members depends on the cross-section geometry. When plastic analysis is used, it is desirable that the members attain their plastic moment capacity with plastic hinge rotation capability without local buckling. Hence, it is necessary that the plate elements of a cross section do not buckle locally due to compressive stresses before plastic hinges are formed.

Example, the compression flange in an I-section is dependent on its width-to-thickness ratio, the local buckling of a cross section before limit state is reached can be avoided by limiting the width-to-thickness ratio of each element of the cross section.

Since the local buckling may affect the bending behaviour of steel sections, steel design specifications define different classes of cross sections, depending on the point at which local buckling occurs during bending. The classification of cross sections subjected to compression due to axial load or bending or shears enables the designer determine whether local buckling influences the section capacity or strength, without calculating their local buckling resistance.

This classification of each element of a cross section subject to compression is based on its width-to-thickness ratio, like other international codes IS: 800-2007 has incorporated the concept of compact and semi-compact sections. The elements of a cross section are generally of constant thickness.

1. Plastic (Class 1) Cross Section

When the width-to-thickness ratio of an element is smaller than that specified in Table 1(Limiting width-to-thickness ratio values as recommended in the IS: 800) , the beam cross section can be fully plasticized and the plastic moment capacity can be developed with plastic hinge rotation capacity to allow redistribution of moments within the section till the failure mechanism is formed, it is referred to as plastic cross section.

2. Compact (Class 2) Cross Section

Class 2 (Compact) Cross-Sections: Such sections can develop plastic moment of resistance, but have inadequate plastic hinge rotation capacity for formation of plastic mechanism, due to local buckling. The sections having width to thickness ratio of plate elements between those specified for class 2.

3. Semi-Compact (Class 3) Cross Section

When the width-to-thickness ratio of a plate in compression is larger than that specified for class 2, the stress at the extreme compression fiber can reach yield or design strength but the elements buckle locally before the complete plasticization of the section can occur. Such a section is referred to as non or semi-compact section. Thus, in this cross section, the plastic moment capacity cannot be developed.

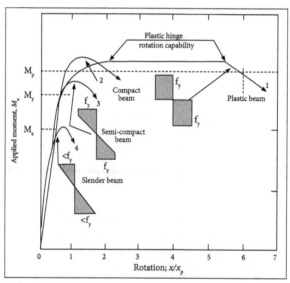

(a) Classification of cross sections based on local buckling influences.

4. Slender (Class 4) Cross Section

If the width-to-thickness ratio of the compression flange is sufficiently large, local buckling of compression flange may occur even before extreme fiber yields; such a section is called a slender section. Thus, in these cross sections, the stress at the extreme compression fiber cannot reach the design or yield strength.

In this type of cross section, it is necessary to make explicit allowance for the effects of local buckling which prevents the development of the elastic capacity in compression and bending. The progressive flexural behaviour of the beams is illustrated in figure (a) in terms of moment rotation characteristics of the cross section. The beam section classified as slender cannot attain the first yield moment because of a premature local or lateral buckling of the web or flange, i.e., the buckling occurs in the elastic range.

The next curve represents the beam as semi-compact in which, extreme fiber stress in the beam section attains yield stress but the section may fail by local buckling before further plastic redistribution of stress can take place towards the neutral axis of the beam. Thus the moment resistance is limited to elastic or yield moment only.

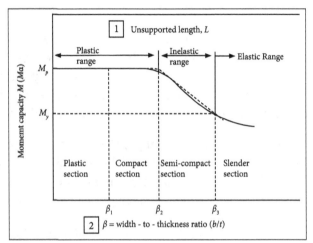

(b) (1) Behaviour of a beam subjected to bending and
(2) Section classification based on width-to-thickness ratio.

The compact cross sections have relatively lower deformation capacity than the plastic sections and the entire section (comprising compression and tension portions) attains yield stress, because of this plastic redistribution of stress, the member attains its plastic moment capacity (M_p) but fails by local buckling before developing plastic hinge rotation capability.

Thus, a compact or class 1 section is deemed suitable for plastic design. The classifications of beam in terms of unsupported length and cross section subject to compression in terms of its width-to-thickness ratio are shown in figure b(1) and b(2), respectively.

The performance of a flexural member in terms of section classification is based on low shear load; the low shear load refers to the factored design shear force that does not exceed $0.6\,V_d$, where V_d is the design shear strength of cross section.

The design bending moment of these classes of beam cross sections are:

- Plastic beam section: $M_d = Z_p f_y$.

- Compact beam section: $M_d = Z_p f_y$.

- Semi-compact beam section: $M_d = M_e = Z_e f_y$.

- Slender beam section: $M_d \leq Z_e f$.

Where Z_e and Z_p are the elastic and plastic section modulus of the beam, respectively.

The design bending strength of a section which is not susceptible to web buckling under shear before yielding, where ($d/t_w < 67\varepsilon$) shall be determined according to Section 8.2.1.2 of the Code.

3.1.1 Lateral Stability of Beams

Lateral instability is a relatively common cause of structural failures, because it is often forgotten or misunderstood. Lateral instability is a result of inadequate lateral stability. Simply defined, lateral stability is the property of an object to develop forces or to have forces imposed upon it that restore it to or maintain its original condition or position. A laterally unstable structure or structural member is able to twist, buckle sideways or fall over.

An example of lateral instability is the twisting or the sideways buckling of a yardstick or a 2×4 set on its narrow edge as a beam when a heavy object is placed on it. Another example of lateral instability is an egg placed on its end or a slender masonry wall. If either of these lean in any direction, their own weight causes them to fall to the side.

Lateral stability is generally provided for a beam, column and steel or wood frame or a wall with lateral bracing. Lateral bracing is the structural component that prevents the beam or column from twisting or buckling sideways or the structural component that prevents the steel frame, wood frame or wall from falling over.

In the case of a beam, the top of the beam is often laterally braced by joists (open web steel joists that look like trusses or I-shaped or C-shaped member) that frame into the side of it or by attaching the floor deck or roof deck to the top of the beam. A column is generally braced in the lateral direction with the beams that frame into its sides.

The lateral bracing for a steel frame, a wood frame or a masonry wall is usually provided intermittently by the floor deck or near its top by the roof deck in combination with diagonal bracing and cross walls. Lateral bracing prevents a steel frame, a wood frame or a masonry wall from falling over under its own weight or during exposure to an impact or wind pressures.

It is unusual for a beam, a column, a steel or wood frame or a wall in a completed structure to be inadequately braced in the lateral direction because they are usually intentionally or inadvertently braced by the adjacent structural components previously mentioned. However, lateral instability is not uncommon during the construction or demolition phases.

During these phases lateral stability must be considered and temporary lateral bracing provided wherever necessary. Lateral bracing is commonly provided during the unstable phases with temporary guy wires, diagonal braces, X-bracing, bracing to a nearby completed structure, etc.

The occurrences of structural failures, injuries and loss of life over the decades due to lateral instability has prompted the preparation and publishing of recommendations, guidelines and/or requirements in industry standards, building codes or OSHA safety rules. Standards and codes providing recommendations and guidelines and building codes providing requirements are available through non-governmental organizations for the construction and engineering industry. Safety rules are available through the labor departments of the state and/or federal governments.

3.2   Lateral Torsional Buckling, Bending and Shear Strength

Lateral Torsional Buckling

Lateral Torsional Buckling in Beams = Lateral Deflection + Torsion

Lateral torsional buckling occurs when an applied load causes both lateral displacement and twisting of a member. This failure is usually seen when a load is applied to an unconstrained, steel I-beam, with the two flanges acting differently, one under compression and the other tension.

'Unconstrained' in this case simply means the flange under compression is free to move laterally and also twist. The buckling will be seen in the compression flange of a simply supported beam. Lateral torsional buckling may occur in an unrestrained beam.

A beam is considered to be unrestrained when its compression flange is free to displace laterally and rotate. When an applied load causes both lateral displacement and twisting of a member lateral torsional buckling has occurred. Figure shows the lateral displacement and twisting experienced by a beam when lateral torsional buckling occurs.

Causes of Lateral Deflection

The applied vertical load results in compression and tension in the flanges of the section. The compression flange tries to deflect laterally away from its original position, whereas the tension flange tries to keep the member straight.

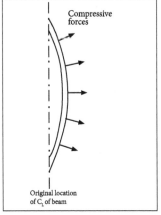

Lateral movement of compression flange.

The lateral movement of the flanges is shown in figure. The lateral bending of the section creates restoring forces that oppose the movement because the section wants to remain straight. These restoring forces are not large enough to stop the section from deflecting laterally, but together with the lateral component of the tensile forces, they determine the buckling resistance of the beam.

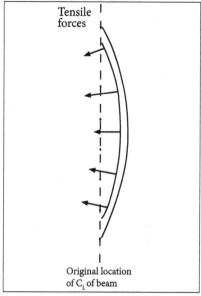

Lateral movement of tension flange.

Bending and Shear Strength

Bending Stresses

Beams are almost always designed on the basis of bending stress and, to a lesser degree, shear stress. A bending stress is not considered to be a simple stress. In other words, it is not load divided by area.

The formula for bending stress, σ_b, is as follows,

$$\sigma_b = My/I$$

Where,

M - Moment acting on beam from moment diagram.

y - Distance from neutral axis to extreme edge of member.

I - Moment of inertia about the axis.

S = I/y, The bending stress formula could be re-written as:

$$\sigma_b = M/S.$$

Where,

S - Section modulus about the axis

Bending stress is distributed through a beam as seen in the diagram below:

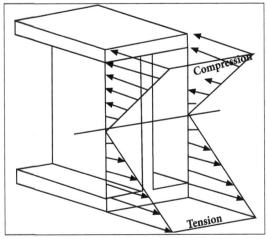

Bending stresses.

A simply-supported beam always has tensile stresses at the bottom of the beam and compressive stresses at the top of the beam.

Shear Stresses

It is easy to imagine vertical shear on a beam that was made up of concrete blocks:

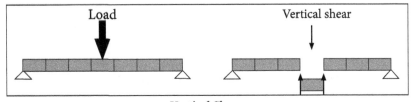

Vertical Shear.

This type of shear is called "transverse" shear and occurs if there is no bending stresses present. The transverse shear stress = V / A. However, almost all real beams have bending stresses present. In this case, beams are more like a deck of cards and bending produces sliding along the horizontal planes at the interfaces of the cards as shown below:

Horizontal Shear.

This type of shear is called "longitudinal" or horizontal shear. The formula used for determining the maximum longitudinal shear stress, f_v, is as follows,

$$f_v = \frac{VQ}{Ib}$$

Where,

V = vertical shear.

Q = First moment of area = A_y.

A = Area of shape above or below the neutral axis.

y = Distance from neutral axis to centroid of area "A".

I = Moment of inertia of shape.

b = width of area "A".

Check for Shear

$V_U = (42.05 \times 8 / 2 + 250 / 2) \times 1.5 = 439.8$ kN

$V_d = (A_v \times f_y w) / (\sqrt{3} * 1.1)$

$= (600 \times 11.8 \times 330) / (\sqrt{3} \times 1.1) = 1226.29 \times 10^3$

$N = 1226.29$ kN

$V_U < 0.6 \times V_d$

< 735.77 kN (Low shear)

Hence OK.

$d / t_w = (600 - 23.6 \times 2) / 11.8 = 46.85 < 67\varepsilon < 58.32$

Hence, design is safe.

3.2.1 Web Buckling and Web Crippling

Web Buckling

At point of concentrated load and at supports, un-stiffened webs of universal beam and compound beams are likely to fail by buckling. The web of the beam is thin and can buckle under reactions and concentrated loads with the web behaving like a short column fixed at the flanges.

The unsupported length between the fillet lines for I sections and also the vertical distance between the flanges or flange angles in built up sections will buckle as a result of reactions or concentrated loads. This is called web buckling.

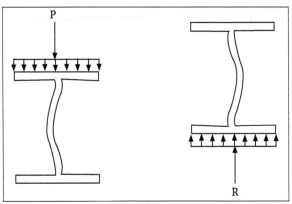

Under Concentrated Load Under support.
Web buckling of beams.

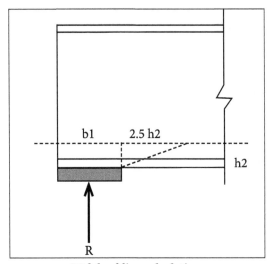

Web buckling calculation.

For safety against web buckling, the resisting force shall be greater than the reaction or the concentrated load. It will be assumed that the reaction or concentrated load is dispersed into the web at 45° as shown in the figure above,

Let Resisting force = F_{wb}

Thickness of web = t_w

Design compressive stress in web = f_{cd}

Width of bearing plate = b_1

Width of dispersion = n_1

$$F_{wb} = (b_1 + n_1)\, t_w\, f_{cd}$$

$\geq$ Reaction, RU

For concentrated loads, the dispersion is on both sides and the resisting force can be expressed as, $F_{wc} = [(b_1 + 2\,n_1)\,t_w\,f_{cd}\,] \geq$ Concentrated load, The design compressive stress f_{cd} is calculated based on an effective slenderness ratio of $0.7\,d\,/\,r_y$,

Where,

d = Clear depth of web between the flanges.

r_y = Radius of gyration about y-y axis and is expressed as,

$$= \sqrt{\left(I_{yy}/\text{area}\right)} = \sqrt{\left[\left[(t_w)^3 /12\right]/t = (t_w)^2 /12\right]}$$

$$k_1/r_y = \left[(0.7\,d)/\sqrt{(t_w)^2 /12}\right] = 2.425 *\ d/t_w$$

Check for Web Buckling

$$k_L /r_y = 2.425 \times d / t_w$$

$$= 2.425 * (600 - 23.6 * 2) / 11.8 = 113.6$$

$$f_{cd} = 97.82\ \text{N} / \text{mm}^2$$

f_{cd} can also be calculated using the equations given in cl.7.1.2.1 as in compression members.

$$F_{wb} = (b_1 + n_1)\,t_w\,f_{cd}$$
$$= (b_1 + 600/200) * 11.8 * 97.82$$

To get,

$$b_1,\ F_{wb} = V_U$$
$$= 439.8 \times 10^3$$
$$b_1 = 81.02\ \text{mm, say 90 mm}$$

90 mm wide bearing plate is provided. (minimum of 75 mm shall be provided).

Web Crippling (or Crimpling)

Web crippling causes local crushing failure of web because of large bearing stresses under reactions at supports or concentrated loads. This happens due to stress concentration because of the bottle neck condition at the junction between flanges and web. It is due to the large localized bearing stress caused by the transfer of compression from relatively wide flange to narrow and thin web.

Web crippling is the crushing failure of the metal at the junction of flange and web. Web crippling causes local buckling of web at the junction of web and flange.

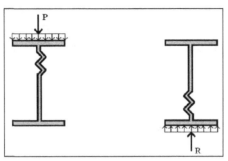

Under Concentrated Load Under support.
Web crippling of beams.

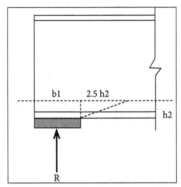

Web crippling calculations.

For safety against web crippling, the resisting force shall be greater than the reaction or the concentrated load. It will be assumed that the reaction or concentrated load is dispersed into the web with a slope of 1 in 2.5 as shown in the figure above.

Let Resisting force = F_{wc}

Thickness of web = t_w

Yield stress in web = f_{yw}

Width of bearing plate = b_1

Width of dispersion = n_2 = 2.5 h_2

Depth of fillet = h_2 (from SP [6])

$$F_{wc} = \left[(b_1 + n_2) \, t_w \, f_{yw} \right] / \gamma_{mo}$$

$\geq$ Reaction, R_U

For concentrated loads, the dispersion is on both sides and the resisting force can be expressed as,

$$F_{wc} = \left[(b_1 + 2\,n_2) \, t_w \, f_{yw} \right] / \gamma_{mo}$$

$\geq$ Concentrated load, W_U

3.2.2 Deflection

When a beam is loaded, it deflects. The amount of maximum deflection depends upon:

- Span.

- Moment of inertia of the section.

- Distribution of the load.

- Modulus of elasticity.

- Support conditions.

Excessive deflection in a floor construction is objectionable not only from architectural point of view but also because of undesirable vibration characteristics.

A very common type of loading is uniformly distributed load (w) along a beam. The maximum deflection, caused at the mid-span of a simply supported beam is given by,

$$\delta = \frac{5}{384} \frac{wL^4}{EI} = \frac{5}{384} \frac{WL^3}{EI} \qquad \qquad ...(1)$$

In general, the maximum, deflection of the beam is given by,

$$\delta = K_t \frac{WL^3}{EI} \qquad \qquad ...(2)$$

Where co-efficient K_1, depends upon the mode of distribution of load. In simply supported beam, for uniformly distributed load, $K_t = \dfrac{5}{384}$ while for concentrated load placed at mid- span, Kl=1/48.

Again, for maximum B.M. at the mid-span,

$$M = K_m WL \qquad \qquad ...(3)$$

Where K_m is a moment coefficient depending upon the distribution of load. For uniformly distributed load, $K_m = 1/8$ while for concentrated load at the mid-span, $K_m = 1/4$.

Again,

$$\frac{M}{I} = \frac{f}{y}$$

or,

$$M = \frac{f}{y} I$$

Taking $f = \sigma_{bc}$ and $y = d/2$ (where d=depth of the beam)

$$M = \frac{2\sigma_{bc}}{d}I \qquad \qquad ...(4)$$

Again,

Hence from (3) and (4),

$$WL = \frac{2\sigma_{bc}I}{K_m d} \qquad \qquad ...(5)$$

Substituting this value of WL in equation (2), we get,

$$\delta = \frac{K_1}{K_m}\frac{2L^2\sigma_{bc}}{Ed}$$

or,

$$\frac{\delta}{L} = \frac{2K_1}{K_m}\frac{L}{d}\frac{\sigma_{bc}}{E} \qquad \qquad ...(6)$$

The above equation is very important since it relates the ratio of maximum deflection and span to the ratio of span to depth.

For uniformly distributed load, where $K_1 = 5/384$ and $K_m = 1/8$, we have,

$$\frac{\delta}{L} = \frac{5}{24}\frac{L}{d}\frac{\sigma_{bc}}{E} \qquad \qquad ...(7)$$

Similarly, for central point load, $K_1 = 1/48$ and $K_m = 1/4$,

$$\frac{\delta}{L} = \frac{1}{6}\frac{L}{d}\frac{\sigma_{bc}}{E} \qquad \qquad ...(8)$$

Limiting deflection: As per IS: 800-1984, the deflection of a member, shall not be such as to impair the strength or efficiency of the structure and lead to damage to finishings. Generally, the maximum deflection should not exceed 1/325 of the span, but this limit may be exceeded in cases where greater deflection could not impair the strengths or efficiency of the structure or lead to damage to finishings. The deflection of a member shall be calculated without considering the impact factor or dynamic effect of the loads causing deflection.

Effective span of beams: The effective span of a beam shall be taken as the length between the centers of the supports, except in cases where the point of application of the reaction is taken as eccentricity to the support, when it shall be permissible to take effective span as the length between the assumed points of application of reaction.

Now taking $\dfrac{\delta}{L} = \dfrac{1}{325}$, we get from equation (6),

$$\frac{1}{325} = \frac{2K_1}{K_m} \frac{L}{d} \frac{\sigma_{bc}}{E}$$

or,

$$\frac{L}{d} = \frac{1}{650} \frac{K_m}{K_1} \frac{E}{\sigma_{bc}}$$ ...(9)

$$\frac{L}{d} = \frac{24}{1625} \frac{E}{\sigma_{bc}} \approx \frac{2954}{\sigma_{bc}}$$ ...(10)

(By taking $E = 2.0 \times 10^5$ N/mm²)

Again,

$$\sigma_{bc} = 0.66 f_y$$

$$\therefore \frac{L}{d} = \frac{4476}{f_y}$$ ...(11)

If $\frac{L}{d} < \frac{4476}{f_y}$, the stress governs the design.

If $\frac{L}{d} > \frac{4456}{f_y}$, the deflection governs the design.

For $f_y = 250$ N/mm²,

$$\frac{L}{d} = \frac{4476}{250} \approx 17.9$$ ...(12)

Check for Deflection

Maximum deflection = $(5wL^4) / (384EI) + (WL^3) / (48EI) = (5 * 42.05 * 8000^4) / (384 * 2 \times 10^5 * 115626.6 \times 10^4) + (250 \times 10^3 * 8000^3) / (48 * 2 \times 10^5 * 115626.6 \times 10^4)$,

= 21.23 mm < L /360 < 8000 / 360 < 22.22 mm

42.05 kN / m = 42.05 N / mm (with actual self-weight of the beam),

$E = 2 \times 10^5$ N / mm² and $I = I_{zz} = I_{xx}$

(from SP - 6, Table I)

3.2.3 Design Procedure

The beams are generally restrained by other parts of the structure, typically in commonly used construction comprising of concrete slabs floor construction wherein the horizontal displacements or rotations of the supporting beams and girders are prevented due to rigid connection between the two components.

The design procedure for a laterally supported beam consists of selecting a suitable section based on the strength and serviceability criteria in terms of its section modulus and overall depth. Then checked for its adequacy in shear capacity, high and low shear web stability, i.e., buckling and crippling of the web.

The steps involved in design are:

1. (a) The estimated different types of service loads are combined by multiplying with appropriate load factors to obtain the design factored load.

(b) An appropriate deflection limit based on the type of finishing materials and function of the structure is decided.

2. The maximum factored or design bending moment M_u and shear force V_u are computed by using the factored load obtained in step 1(a).

3. The trial plastic section modulus about z-axis for the beam is determined from,

$$Z_{p,z} = \frac{M_u}{\left(f_y / \gamma_{mo}\right)} \qquad \qquad ...(1)$$

The trial elastic section modulus about Z—axis for semi-compact section within four per cent of the exact value (conservative value) for plastic and compact sections is given by,

$$Z_{e,z} = \frac{M_u}{f_y} \qquad \qquad ...(2)$$

4. Based on the required plastic section modulus determined in step 3 and predetermined depth of the section of step 1(b), the lightest section is selected from the steel table as a trial section and its properties are recorded.

5. The class of section is ascertained by computing the dimensionless parameters b/t_f and d/t_w. Depending upon the specified limits, these parameters will determine the class of the section, i.e., whether it is plastic or compact or semi-compact or slender sections.

6. The trial section is checked for its shear capacity. For the adequacy of the trial section in shear its design shear strength V_d should not be less than the design or factored shear force V_u.

7. Based on design shear strength V_d computed in step 6, the shear condition to be used in computation of design bending strength is ascertained; when $0.6\,V_d \geq V_u$, low shear condition and $0.6\,V_d < V_u$ high shear condition shall apply.

8. For adequacy of trial section in bending, the design bending strength M_d calculated with shear condition ascertained in step 6 should be more than the design bending moment, M_u.

(a) The shear force does not have any influence on bending moment for values shear up to 0.6 V_d, called low shear condition, where V_d is the design shear force. For low shear condition,

$$M_d = \beta_b Z_p \left(f_y / \gamma_{mo} \right) \leq 1.2\, Z_e \left(f_y / \gamma_{mo} \right) \text{ for cantilever} \qquad \qquad ...(3)$$

Where,

$\beta_b = 1.0$, for plastic and compact sections and $\beta_b = Z_e/Z_p$, for a semi-compact section.

(b) The case when design (factored) shear force V_u exceeds 0.6 V_d, it is called the high shear condition and bending strength is calculated as follows:

i. Class 1 and 2 plastic and compact sections For high shear condition bending strength of plastic and compact sections is given by,

$$M_{d,v} = M_d - \beta_b \left(M_d - M_{jd} \right) \leq 1.2\, Z_e \left(f_y / \gamma_{mo} \right) \qquad \qquad ...(4)$$

Where,

$$M_{jd} = \left[Z_{p-} \left(t_w h^2 / 4 \right) \left(f_y / \gamma_{mo} \right) \right]$$

And $\beta = (2 V_u / V_p - 1)^2$ in which $V_p = 0.6(h\, t_w\, f_y)$ represents design shear strength as governed by web yielding or web buckling.

ii. Class 3 Semi-compact section Most of the standard rolled I- and channel shapes are compact; the remaining are semi-compact because of flange width-thickness ratio. For a semi-compact section the moment capacity is given by,

$$M_d = Z_e \left(f_y / \gamma_{mo} \right) \geq M_u \qquad \qquad ...(5)$$

$$M_d = f_y \left(Z_e - \beta\, Z_{p,y} / 1.5 \right) \qquad \qquad ...(6)$$

iii. Class 4 slender sections As per Euro code,

$$M_d = f_y \left(Z_{e,eff} - \beta\, Z_{p,y} / 1.5 \right) \qquad \qquad ...(7)$$

The adequacy of resistance of the trial section against web buckling is checked. In case of low shear condition, the webs without stiffeners and having $d/t_w \leq 67\varepsilon$ are considered to be safe in web buckling and the shear strength of the web is governed by the plastic shear resistance.

However, in case of high shear condition, the web should be checked for buckling. Web buckling strength should not be less than the design shear force.

The web buckling strength of the section

$$= A_b f_{d,c} = \left[t_w \times \left(b_1 + 2n_2 \right) \right] f_{d,c} \qquad\qquad ...(8)$$

Where,

A_b = Bearing area of the web at the neutral axis of the beam $= t_w \left(b_1 + 2n_2 \right)$.

$f_{d,c}$ = The design compressive stress.

9. For the adequacy of trial section to resist the web buckling safely, $F_w \geq V_u$ Web bearing (crippling) strength,

$$F_w = A_e \left(f_{y,w} / \gamma_{mo} \right) = \left[\{ b_1 + 2.5(t_f + R_1) \} t_w \right] \left(f_{y,w} / \gamma_{mo} \right) \qquad\qquad ...(9)$$

$f_{y,w}$ = Yield stress of the web of the beam section.

10. The maximum deflection of trial section is computed for the service loads and to satisfy the serviceability criterion, the computed deflection should be less than the permissible deflection (L/300). However, when the selection of trial section in step 4 is based on predetermined depth of the section of step 1(b), then this test is not required.

Design Steps for Laterally Unsupported Beams

Laterally Unsupported Beam

When the compression flange is not supported it has a tendency to bend in the lateral dissection with twisting. This is called laterally unsupported beam.

Example: Flange Cleats.

The design of laterally unsupported beams consists of selecting a section based on the plastic section modulus and checking for its shear capacity, deflection, web buckling and web crippling. Most of the equations are available in IS 800: 2007.

All the steps given in design of laterally supported beams shall be used here. Also, The plastic section modulus required is increased by 25 - 50%. The design bending strength is calculated using the appropriate provisions in the code for lateral supported beams. Other checks like deflection, shear and local criteria will be same.

Design Steps for Laterally Supported Beams

The design of laterally supported beams consists of selecting a section based on the plastic section modulus and checking for its shear capacity, deflection, web buck ling and web crippling. Most of the equations are available in IS 800 : 2007. The steps are,

- The maximum BM and SF at collapse is calculated based on the service loads (characteristic loads) and the span of the beam.

  Factored load at collapse = 1.5 X characteristic loads.

- The plastic section modulus, ZPZ is calculated using,

  $Z_{PZ} = M_U / (f_Y / 1.1)$

  M_U = Maximum BM

  f_Y = Yield stress of the given grade of steel

  1.1 = Partial safety factor in yielding

- A trial section having the appropriate plastic section modulus is adopted using IS 800 or SP (6) depending upon the type of section required. The section shall be plastic or compact section.

- The beam shall be checked for shear lag and design bending strength as given in cl. 8.2.1.5 and 8.2.1.2 (pp -53).

- The beam is checked for deflection using appropriate formula depending on the type of loadings.

- The section is checked for shear as given in cl. 8.4, 8.4.1 and 8.4.1.1. If VU $\geq$ 0.6 Vd it is a case of high shear or otherwise low shear. For high shear, the design bending strength is calculated from cl. 9.2

- The section is checked for web buckling and crippling using appropriate formula.

Problems

1. Let us design a simply supported beam of 10 m effective span carrying a total factored load of 60 kN/m. The depth of beam should not exceed 500 mm. The compression flange of the beam is laterally supported by floor construction. Assume stiff end bearing is 75 mm.

Solution:

Given:

L = 10 m = 10000 mm, w = 60 kN/m

Trial Section

Maximum BM,

$$M = \frac{wL^2}{8} = \frac{60 \times 10^2}{8} = 750 \text{ kN/m}$$

$\therefore Z_p$ required,

$$\frac{M\gamma_{mo}}{f_y} = \frac{750 \times 10^6 \times 1.1}{250} = 3300 \times 10^3 \text{ mm}^3$$

Since depth restricted is 500 mm, select ISMB 450 and suitable plates over flanges.

Z_p of ISMB 450 = 1553.4 × 10³

Z_p to be provided by cover plates

= (3300 − 1553.4) × 10³

= 1746.6 × 10³ mm³

If A_p is the area of cover plate on each side tensile force and compressive forces developed at the time of hinge formation = $A_p f_y$.

If the distance between the two plates is 'd', plastic moment resisted = $A_p f_y d$.

Hence the additional Z_p provided by the cover plates may be taken as,

$$Z_p \text{ of plates} = \frac{A_p f_y d}{f_y} \times \frac{1}{\gamma_{mo}} = \frac{A_{pd}}{1.1}$$

$$\therefore \frac{A_{pd}}{1.1} = 1746.6 \times 10^3$$

Taking d = 450 + t = 450 mm.

We get,

$$A_p = \frac{1746.6 \times 10^3 \times 1.1}{450} = 4269.5 \text{ mm}^2$$

Provide 220 × 20 mm plates on either side.

Check for Shear

$$V_d \frac{f_y}{\sqrt{3}} \times \frac{1}{\gamma_{mo}} \times h \times t_w = \frac{250}{\sqrt{3}} \times \frac{1}{1.1} \times 450 \times 9.4$$

$$= 555.044 \times 10^3 \text{ N}$$

$$V = 60 \times \frac{10}{2} = 300 \text{kN}$$

Check for Moment Capacity

Section classification: Outstanding element of compression flange,

$$\frac{b}{t_b} = \frac{150/2}{17.4} = 4.3 < 9.4 \text{ E}$$

$$\frac{d}{t_w} = \frac{450 - 2(19.4 + 15)}{9.4} = 40.9 < 84 \text{ E}$$

Hence plastic section,

$$H_d = 1.0 \times \frac{Z_p f_y}{\gamma_{mo}} < 1.2 \, Z_e \, f_{hy} \frac{1}{\gamma_{mo}}$$

$$\frac{Z_p f_y}{\gamma_{mo}} = \left[1553.4 \times 10^3 + 220 \times 20 (450 + 20) \right] \times \frac{250}{1.1}$$

$$= 823.0455 \times 10^6 \text{ N} - \text{mm}$$

$$I_{zz} = 789.888 \times 10^6 \text{mm}^4$$

$$Z_e = \frac{789.88 \times 10^6}{255 + 20} = 3224.033 \text{ mmm}^4$$

$$M_d = 823.043 \times 10^6 > \text{H.}$$

Hence, design is safe.

Check for Deflection

Working load,

$$= \frac{60}{1.5} \text{ kN/m} = 40 \text{ kN/m} = 40 \text{ N/mm}$$

$$\delta = \frac{\delta \omega L^4}{384 \, E I_{zz}} = \frac{5 \times 40 \times (10000)^4}{384 \times 2 \times 10^5 \times 789.884 \times 10^6} = 32.97 \text{ mm}$$

Permissible, if elastic cladding is assured.

(Table of IS 800), $\dfrac{L}{240} = \dfrac{10,000}{240} = 41.67$ mm

Hence safe.

Check for Web Buckling

$$h = 450 \text{ mm.}$$

$$\therefore \lambda = 2.5\frac{h}{t_w} = 2.5 \times \frac{450}{9.4} = 119.68$$

$$f_{cd} \quad 94.6 - \frac{9.68}{10}(94.6 - 83.7) = 84.05 \text{ N/mm}^2$$

$$F_{cdw} = (b_1 + n_1)t_w f_{cd} = \left(75 + \frac{450}{2}\right)9.4 \times 84.05$$

$$= 237.021 \times 10^3 \text{N} < 300 \text{ N.}$$

Hence, for increased effective bearing length, Provide b = 175mm, then,

$$F_{cdw} = \left(175 + \frac{450}{2}\right) \times 0.4 \times 84.05$$

$$= 316.028 \times 10^3 > 300 \text{ kN.}$$

$\therefore$ Safe.

Check for Web Crippling

$$F_w = (b_1 + 2.5t_f)f_y\frac{1}{\gamma_{mo}}t_w$$

$$= \left[175 + 2.5 \times (17.4 + 20)\right]250 \times \frac{1}{1.1} \times 9.4$$

$$= 573.614 \times 10^3 \text{ N} > 300 \text{ kN.}$$

Hence safe.

Design of Connection between Flange Plates and Flange

Bolts/welds joining the plates and flange are to be designed for the horizontal shear at that level. Shear stress at the level of plates and flanges is given by,

$$= \frac{F}{bI_{zz}}(\bar{a}_y) = \frac{300 \times 10^3}{150 \times 789.884 \times 10^6}(220 \times 20 \times 225)$$

$$= 2.506 \text{ N/mm}^2$$

2. Design a simply supported laterally restrained beam of effective span 4m carrying

a factored point load of intensity 50kN at the midspan. Let us design an appropriate section using Fe410 grade steel.

Solution:

Given:

Span = 4m

Load of intensity = 50kN

Step 1: Calculation of BM and Choosing a Trial Section:

Maximum B.M,

$$M = \frac{\omega I}{4} = \frac{50 \times 4}{4} = 50 \text{ kN.m}$$

$$M = 50 \times 10^6 \text{ N.mm}$$

Maximum Shear,

$$= \frac{\omega I}{2}$$

$$= \frac{50 \times 4}{4}$$

$$= 100 \text{ kN.m}$$

$$Z_p \text{ required} = \frac{M\gamma_{m_o}}{f_y} = \frac{50 \times 10^6 \times 1.1}{250} = 2.2 \times 10^6 \text{ mm}^3$$

ISMB 550 with suitable plates over the flanges.

Step 2: Choosing Cover Plates:

Z_p of ISMB 400 = 11.76, 20 × 10³mm³

Z_p to be provided by the cover plate.

$$= 2.20 \times 10^6 - 1.176 \times 10^6$$

$$= 1.0 \times 10^6 \text{mm}^3$$

Let A_p be the area of cover plate on each side of the flange.

$$E = 2 \times 10^5 \text{ N/mm}^2$$

$$Z_R = 2359.8 \text{ cm}^3 = 2359 \times 10^3 \text{ mm}^3$$

$Z_p = 2711.08 \text{ cm}^3 = 2711.98 \times 10^3 \text{ mm}^3$

$D = 550 \text{ mm}$

$b_f = 190 \text{ mm}$

$t_f = 19.3 \text{ mm}$

$t_\omega = 11.2 \text{ mm}$

$f_y = 250 \text{ N/mm}^2 \quad r_1 = 18 \text{mm}$

$\varepsilon = \sqrt{\dfrac{250}{f_y}} = 1$

$b = \dfrac{b_f}{2} = \dfrac{190}{2} = 95 \text{mm}$

$\dfrac{b}{t_f} = \dfrac{95}{19.3} = 4.92 < 9.4\varepsilon$

∴ Plastic Section.

$d = D - 2t_f - 2r_1$

$= 550 - 2[19 - 3] - 2[18]$

$d = 475.4 \text{ mm}$

$\dfrac{d}{t\omega} = \dfrac{475.4}{11.2} = 42.45 \rightarrow 84\varepsilon$

∴ Plastic Section.

∴ Section is classified as plastic section.

Design Shear Strength of the Section

[as per Cl.8.4 of IS 800:2007].

$V = 100 \text{kN}$

$V_d = \dfrac{V_n}{\gamma_{m_0}}$

$V_n = \dfrac{A_v f_{y\omega}}{\sqrt{3}}$

$$= \frac{[ht_\omega]f_y}{\sqrt{3}} = \frac{550 \times 11.2 \times 250}{\sqrt{3}}$$

$$V_n = 889 \text{ kN}$$

$$V_d = \frac{889}{1.1} = 808 > 100 \text{ kN}$$

Hence Safe.

Design Bending Strength of the Section

The section is plastic and laterally supported,

$$= \frac{\beta_b Z_p f_y}{\gamma_{m_o}} \le \frac{1.2 Z_e f_y}{\gamma_{m_o}}$$

$\beta_b = 1$ for Plastic Section

$Z_p = 2711.98 \times 10^3 \text{mm}^3$

$Z_e = 2359 \times 10^3 \text{ mm}^3$

$$= \frac{1 \times 2711.98 \times 10^3 \times 250}{1.1} \le \frac{1.2 \times 2359 \times 10^3 \times 250}{1.1}$$

$$= \frac{616 \text{ kN.m} \le 643.3 \text{ kNm}}{\text{Series sake}}$$

$$= 1.0 \times 10^6 \text{ mm}^2$$

Let A_p be the area of cover plate on each side of the flange.

Limiting Deflection

$$y = \frac{\omega L^3}{48EI} \le \frac{I}{300}$$

$$= \frac{50 \times 10^3 \times 4000^3}{48 \times 2 \times 10^5 \times 64893.34 \times 10^4} \le \frac{4000}{300}$$

$$y = 0.513 \le 13.33 \text{ mm}$$

Hence Safe

3. Let us design a laterally unsupported beam of 4m effective span, carrying a factored

bending moment of 350 kNm and factored shear force of 100 kN. Use Fe410 grade steel.

Solution:

Given:

Design Laterally Unsupported Beam

Effective Span = 4m.

Factored B.M = 350kN.m.

Factored S.F = 100kN.

Use Fe410 grade steel.

$\Rightarrow$ Choose ISMB i.e., 550

$Z_p = 2711.9 \times 103\ 10/mm3/Z_e = 2359 \times 103\ mm3$

$D = h = 550$ mm

$b_f = 190$ mm

$t_f = 19.3$ mm

$t_w = 11.2$ mm

$\gamma_1 = 221.6$ mm

$f_y = 410$ N/mm²

$\varepsilon = \sqrt{250/f_y} = 0.78$

$$b = \frac{b_f}{2} = \frac{190}{2} = 95\,mm$$

$$\frac{b}{t_f} = \frac{95}{19.3} = 4.92 \le 9.4\varepsilon$$

[Plastic Section]

$d = D - 2[t_f + \gamma_1]$

$= 550 - 2\,[19.3 + 18]$

$d = 475.4$ mm

$$\frac{d}{t_w} = \frac{1475.4}{11.2} = 42.96 \le 84\varepsilon\ \text{[Plastic Section]}.$$

f_{crb}:

$\qquad$ KL = 4000mm

$$\frac{K^2}{\gamma_{xx}} = \frac{4000}{221.6} = 18.05$$

$$\frac{b}{t_f} = \frac{550}{19.3} = 28.20$$

Refer Table 14 of IS 800, for (KL/r) = 18.05 & $\dfrac{h}{t_f}$ = 28.2

$\qquad$ f_{crb} = 7.4 N/mm²

f_{bd}:

$\qquad$ αL_T = 0.21 for rolled section.

Refer Table 13.9 for αL_T = 0.21, f_y = 40 N/mm²

$\qquad$ f_{bd} = 374.1 N/mm²

Design bending strength,

$\qquad$ $M_d = \beta_b \; z_p \; f_{bd}$

$\qquad$ β_b = 1 [for Plastic Section]

$\qquad$ $Z_p = 2711.9 \times 10^3$ mm³

$\qquad$ $M_d = 1 \times 2711.9 \times 10^3 \times 216.9$

$\qquad$ $M_d = 588.23 \times 10^6$ N/mm & 350×15^6 N.mm

Design Shear Strength of the Section

$$V_d = \frac{V_A}{\gamma_{m_y}}, V_n = \frac{A_V \, f_{yw}}{\sqrt{3}}$$

$$V_n = \frac{h \, t_w \, f_{yw}}{\sqrt{3}} = \frac{550 \times 11.2 \times 410}{\sqrt{3}}$$

$\qquad$ V_n = 1458.15 kN

$$V_d = \frac{1458.15}{1.1} = 1325.5 \text{ kN} > 100 \text{kN}$$

Hence Safe.

4. A laterally supported beam having an affective span of 8 m consists of ISMB 550 at 103.7 kg/m and covet plate of 250 mm × 16 mm connected to each flange by 20 mm diameter rivets. Let us determine the safe UDL which the beam can carry in addition of its own weight.

Solution:

Given:

Effective Span = 8 m

ISMB 550 at 103.7 kg/m

Cover Plate 250 mm × 16 mm

Connect to each flange 20 mm dia rivets.

Section Properties

$D = 550$ mm

$b_f = 190$ mm

$t_f = 19.3$ mm

$t_w = 11.2$ mm

$r_1 = 22.16$ mm

$f_y = 250$ MPa

$Z_p = 2711.98 \times 10^3 \text{mm}^3$

$Z_e = 2359.80 \times 10^3 \text{mm}^3$

Classification of Section,

$$b = \frac{b_f}{2} = \frac{190}{2} = 95 \text{mm}$$

$$\frac{b}{t_f} = \frac{95}{19.3} = 4.9 < 9.4\varepsilon \quad \text{[Plastic Section]}$$

$$d = D - 2[t_f + r_1]$$

$$= 550 - 2[19.3 + 22.16]$$

$$d = 467.08 \text{ mm}$$

$$\frac{d}{t_w} = \frac{467.08}{11.2} = 41.7 \text{ mm} < 84 \ \varepsilon \ \text{[Plastic Section]}.$$

To find f_{crb},

$$KL = L = 8000 \text{ mm}$$

$$\frac{KL}{\gamma_{xx}} = \frac{8000}{221.6} = 36.1$$

$$\frac{h}{t_f} = \frac{550}{19.3} = 28.49$$

$$\frac{kL}{r} = 36.1 \ \& \ \frac{h}{t_f} = 28.49$$

$$f_{crb} = 1680.1 \text{ N/mm}^2$$

To find f_{bd},

$$\alpha \ L_T = 0.21 \text{ for rolled section.}$$

By referring for,

$$\alpha \ L_T = 0.21 \ \& \ f_y = 250 \text{ MPa}$$

$$f_{crb} = 1680.1 \text{ N/mm}^2$$

$$f_{bd} = 224 \text{ MPa}$$

Design Bending Strength,

$$M_d = \frac{\beta_b \, Z_p \, f_y}{\gamma_{m_0}} \leq \frac{1.2 Z_e \, f_y}{\gamma_{m_0}}$$

$$\beta_0 = 1 \text{ for plastic section.}$$

$$M_d = \frac{1 \times 250 \times 2712 \times 10^3}{1.1} \leq \frac{1.2 \times 2359.8 \times 10^3 \times 250}{1.1}$$

$$= 616363636.4 \leq 643581818.2 \text{ N}$$

Factored udl for the Section:

$$M = \frac{\omega l^2}{8}$$

$$= 616363636.4 = \frac{\omega \times 8000^2}{8}$$

$$\omega = 077.05 \text{ N /mm Self Weight.}$$

Design Shear Strength

$$V_d = \frac{V_n}{\gamma_{m_0}}$$

$$V_n = V_p = \frac{A_v d_{yw}}{\sqrt{3}}$$

$$V_n = \frac{[550 \times 11.2]250}{\sqrt{3}} = 8.89119\,N$$

$$V_d = \frac{889119}{1.1} = 808290\,N$$

Factored udl on Design Shear,

$$\frac{\omega l}{2} = V_d$$

$$808290 = \frac{\omega \times 8000}{2}$$

$\omega = 202.07\text{N/mm} + $ Self Weight

$\therefore$ Permissible udl 77.05 N/mm = 77.05 kN/m + Self-Weight

5. Let us design a simply supported beam of effective span 12 m to carry a load factor 70 kN/m the depth of beam is restricted to 500 mm.

Solution

Given:

Effective Span = 1200

Load factor u dl on beam = 70 kN/m

Step 1: Calculation of B.M and choosing a trial section.

Maximum BM,

$$\frac{Wl^2}{8} = \frac{70(12)^2}{8} = 1260\,kN.m$$

$$= 1260 \times 10^6 \text{ N.mm}$$

$$Z_p \text{ required} = \frac{M\gamma_{mo}}{f_y} = \frac{1260 \times 10^6 (1.1)}{250}$$

$$= 5.511 \times 10^6 \text{ mm}^3$$

Step 2: Choosing cover plate = 1176.2×10^3 mm³

$$\therefore Z_p \text{ cover plate} = (5.544 \times 10^6 - 1176.2 \times 10^3)$$

$$= 4.368 \times 10^6 \text{ mm}^3$$

(Area of Plate) $\dfrac{f_y}{\gamma_{mo}}$ (Distance between the plate)

$$= \frac{A_p f_y}{\gamma_{mo}}(400)$$

$$M = f_y \cdot Z_p$$

$$Z_p = \frac{A_p f_y (4000)}{\gamma_{mo} f_y} = \frac{A_P (400)}{1.1}$$

$$= 4.368 \times 10^6 \text{ mm}^3$$

$$M_d = B_b \frac{Z_p f_y}{\gamma_{mo}}$$

$$= (1.0)\left[\frac{Z_P \cdot f_y}{\gamma_{mo}}\right] (Z_p \text{ of section} + Z_p \text{ plate})$$

$M_d = 1176.2 \times 10^3 + (320 \times 40)(400 + 40)] .(250/1.1)$

$M_d = 1547.318 \times 10^6$ N/mm

$Z_{fd} = [Z_p \text{ I section} + Z_p \text{ of plate} - A_w /yw]$

$$= 1176.2 \times 103 + (320 \times 40)(400 * 410) - \left(400 \times 8.9 \times \frac{400}{4}\right)$$

$$= S [1176.2 \times 10^3 + 5632 \times 10^3 - 356 \times 10^3]$$

$$= 6452.2 \times 10^3 \text{ mm}^3$$

$M_{dv} = M_d - \beta (M_d - M_{fd})$

$$= 1547.318 \times 10^6 - 0.6371 (1547.31 \times 10^6 - 1466.44 \times 10^6)$$

$$= 1547.318 \times 10^6 - 51.53 \times 10^6$$

$\Rightarrow 1496 \times 10^6$ Nmm

Working load = $(70/1.5) = 46.67$ kN/m

$$4.368 \times 10^6 = \frac{A_P (400)}{1.1}$$

$A_p = 12012$ mm²

Step 3: Check for shear

$$V_d = \frac{(f_y / \sqrt{3})}{\gamma_{mo}} ht_w = \frac{(250\sqrt{3})}{1.10} \times 100 \times 8.9 = 467143 \text{ N.}$$

$0.6 V_d = 0.6 (467143) = 280286$ N

$$S.F = \frac{W_1}{2} = \frac{70000}{2}(12) = 420 \times 10^3 \text{ N}$$

Step 4: Section classification

$$\frac{b}{t_f} = \frac{140}{6} = 8.75 < 29.3$$

$$\frac{d}{t_w} = 400 - \frac{2(t_f + r_1)}{t_w}$$

$$= \frac{400 - 2(16.0 + 14)}{8.9}$$

$$= 38.2024$$

Step 5: Check for B.M

$$M_{dv} = M_d - \beta (M_d - M_{fd})$$

$$\beta = \left(\frac{2V}{V_d - 1} \right)^2$$

$$= \left[\frac{2(420 \times 10^3)}{467143} - 1 \right]^2$$

$$= 0.6371$$

$$S = \frac{5WS^4}{384 EI_{zz}} = \frac{5 \times 46.67 \times (12000)^4}{384(2 \times 10^5) I_{zz}}$$

$I_{xx} = I_{zz}$ for ISMB 400 + I_{zz} for plate

$$= 20458.4 \times 104 + 2 \times AP \left(\frac{d}{2}\right)^2$$

$$= 20458.4 \times 4 \times 104 + 2 \left[(320 \times 40) \left(\frac{400}{2} + \frac{40}{2}\right)^2 \right]$$

$$= 144.36 \times 10^7 \text{ mm}^4$$

$$S = \frac{5 \times 46.67 \times (120000)^2}{384(2 \times 10^5) \times (144.36 \times 10^7)}$$

$$= 43.64 \text{ mm}$$

$$= \frac{L}{240}$$

$$= 50 \text{ mm} > 43.64 \text{ mm}$$

Step 6: Check for web buckling.

$$d = 2.5 \frac{h}{t_w} = 2.5 \left(\frac{400}{8.9}\right)$$

$$= 112.36 \text{ mm}$$

$$f_{cd} = 94.6 \frac{(94.6 - 83.7)}{10} (2.36)$$

$$= 92.03 \text{ MPa}$$

$$E_{cdw} = (b_1 + n_2) \, t_w \, (f_{cd})$$

$$b_1 = 70 \text{ mm}$$

$$n_2 = \frac{400}{2} = 200 \text{ mm}$$

$t_w = 8.9 \text{ mm}$

$f_{cd} = 92.03 \text{ MPa}$.

$F_{cdw} = (70 + 200) \, (8.9) \, (92.03) = 221148 \text{ N}$.

$$= 220 \times 10^3 \text{ N}$$

$F_{cdw} = (350 + 200) \, (8.9) \, (92.093)$

$$= 450.795 \times 10^3 \text{ N} = 420 \times 10^3 \text{ N}$$

Check for Web Crippling,

$$F_w = \frac{(b_1 + n_2)(t_w)(f_y)}{\gamma_{mo}}$$

b_1 = 350 mm

n_2 = 2.5 (root radius + flange thickness)

  = 2.5 (14 + 16) = 75 mm

t_w = 8.9 m

f_y = 250 MPa

γ_{mo} = 1.1

$$F_w = \frac{(350.175)(8.9)(250)}{1.1} = 259.65 \times 10^3 \text{ N} > 420 \times 10^3 \text{ N}$$

6. A beam is simply supported over a span of 6 m. It supports one Iron beam at mid span exerting 90 kN. Let us design the beam with ISWB section with flange plates.

Solution:

Given:

Span l = 6m

Approximate deal load of the beam,

$$= \frac{\text{L.L.} \times \text{Span}}{300} = \frac{30 \times 6}{300}$$

= 0.6 kN/m

Total load = 30 + 0.6 = 30.60 kN/m

Minimum B.M = M $= \frac{\omega l^2}{8} = \frac{30.6 \times 6^2}{8} = 137.7 \text{ kN / m}$

Minimum S.F = S $= \frac{30.6 \times 6^2}{2} = 91.8 \text{ kW}$

Permissible bending stress = 0.66 × 250 = 165 N/mm²

Permissible average shear stress = 0.4 × 250 = 100 N/mm²

Section modulus required Z $= \frac{M}{\sigma_{bc}} = \frac{137.7 \times 10^6}{16^5} \text{mm}^2$

$Z = 834.5 \times 10^3$ mm^3

Provide ISWB at 56.9 Ks/m

Properties of the section overall depth 350 mm,

$t_\omega = 8$ mm

$I_{xx} = 15,321.7 \times 10^4$ mm^4

$Z = 887 \times 10^3$ m^3

$h_2 = 27.25$ mm

$h_1 = 255.5$ mm

Check for Shear,

Induced average shear stress $= \dfrac{S}{dt_\omega} = \dfrac{91.8 \times 10^3}{350 \times 8}$ N/mm^2

This is less than permissible shear stress.

Check for Deflection,

Maximum deflection at the centre $= \dfrac{5}{384} \dfrac{\omega l^4}{El}$

$= \dfrac{5}{384} \dfrac{30.6 \times 6^4 \times 10^9}{200 \times 15321.7 \times 7 \times 10^4} = 16.85$ mn

Permissible deflection $= \dfrac{\text{Span}}{325} = \dfrac{6000}{325} = 18.46$ mm

Since the maximum deflection is less than permissible deflection, the design is safe.

Check for Crippling Web at Support

Assume bearing of 300 mm,

Induced bearing stress $= \dfrac{R}{t\left(a + h^2\sqrt{3}\right)} = \dfrac{91.8 \times 10^3}{8/300 + 27.25\sqrt{3})}$

$= 33$ N/mm^2 187.5 N/mm^2

Check for Buckling of the Web

Slenderness ratio from used buckling,

$\dfrac{h_1}{t}\sqrt{3} = \dfrac{295.5}{8}\sqrt{3} = 64$

Safe compressive stress corresponding to above ratio,

$\sigma_{ac} = 117$ N/mm^2

$B = 300 + (h/2) = 300 + \dfrac{350}{} = 475$ mm

Load bearing capacity of the web = σ_{ac} t B = 117 × 8 × 475 = 444.6 kN.

But the end reaction is only 91.8 kN.

3.3   Plate   Girders: Various   Elements   and   Design   of Components

Main Plate Girders

The design criterion for main girders as used in buildings, are discussed here on Plate Girders. Generally, the main girders require web stiffening (either transverse or both transverse and longitudinal) to increase efficiency. Sometimes variations of bending moments in main girders may require variations in flange thickness to obtain economical design. This may be accomplished either by welding additional cover plates or by using thicker flange plate in the region of larger moment.

In very long continuous spans (span> 50 m) variable depth plate girders may be more economical. Initial design of main plate girder is generally based on experience or thumb rules such as those given below. Such rules also give a good estimate of dead load of the bridge structure to be designed. For highway and railway bridges, indicative range of values for various overall dimension of the main girders are given below,

Overall depth, D: $l/18 \leq D \leq l/12$ (Highway bridges)

$l/10 \leq D \leq l/7$ (Railway bridges)

Flange width, 2b: $D/4 \leq 2b \leq D/3$

Flange thickness, T: $b/12 \leq T \leq b/5$

Web thickness, t: $t \approx D/125$

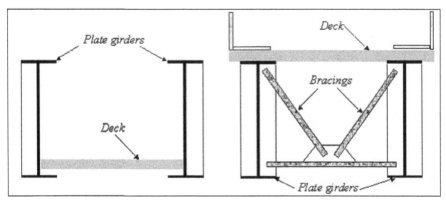

(a) Half through type plate grinder bridge
(b) Deck type grinder bridge Common types of plate girder bridge.

Where, l is the length between points of zero moment. The detailed design process to maximize girder efficiency satisfying strength, stability, stiffness, fatigue or dynamic criteria, as relevant, can be then carried out. Recent developments in optimum design methods allow direct design of girder bridges, considering minimization of weight/ cost.

Plate Girder Design

The load effects (such as bending moment and shear force) are to be found using individual and un-factored load cases. Based on these the summation of load effects due to different load combinations for various load factors are obtained. Since bridges are subjected to cyclic loading and hence are vulnerable to fatigue, redistribution of forces due to plastic mechanism formation is not permitted.

The design is made based on limit State of collapse for the material used considering the following:

- Shape limitation based on local buckling.
- Lateral torsional buckling.
- Design of Steel Structures.
- Web buckling.
- Interaction of bending and shear.
- Fatigue effect.

Design Consideration

- Design Limit States, Steel girder bridges shall be designed to meet the requirements for all applicable limit states specified by AASHTO and the California Amendments such as Strength I, Strength II, Service II, Fatigue I and II are extreme events. Construct ability must be considered.

- Design requirements perform the following design portions for an interior plate girder in accordance with the AASHTO LRFD Bridge Design Specifications with the California Amendments.

Select Girder Layout and Sections

- Perform Load and Structural Analysis.
- Calculate Live Load Distribution Factors.
- Determine Load and Resistance Factors and Load Combinations.

- Calculate Factored Moments and Shears – Strength Limit States.

- Calculate Factored Moments and Shears – Fatigue Limit States.

- Calculate Factored Moments – Service Limit State II.

- Design Composite Section in Positive Moment Region at 0.5 Point of Span 2.

- Design Non composite Section in Negative Moment Region at Bent 3.

- Design Shear Connectors for Span 2.

- Design Bearing Stiffeners at Bent 3.

- Design Intermediate Cross Frames.

- Design Bolted Field Splices.

- Calculate Camber and Plot Camber Diagram.

- Identify and Designate Steel Bridge Members and Components.

Plate Girder and Its Uses

Building up I-section with two flanges plates connected to web plate of required depth. This type of I-beams are known as "Plate Girder".

Used in road or railway bridges, to carry crane beam, plate girder are used.

The Necessity of Curtailment of Flange Plate in the Plate Girder

The value of bending moment is maximum at the centre and it gradually reduces to zero in case of simply supported beams. Hence larger cross section is required at the mid span.

The cross section could be reduced to just meet the required bending moment which could help in reducing the cost of the plate girder.

Factors Governing flange Curtailment in Plate Girders

Often the plate girder is subjected to loading then the maximum bending moment at one section usually when the plate girder is simply supported at the ends and subjected to the uniformly distributed load (Udl) then the maximum BM occurs at the flange of plate girder is designed to resist the minimum BM.

The flange area designed to resist the maximum BM is not required at other section. The flange area reduced as the BM decreases. The reduction in flange area is done by curtailing the flange plate.

Stiffeners are normally classified as vertical and horizontal stiffeners. Vertical stiffeners are used to transmit the concentrated load to the web plate. Horizontal stiffeners are provided in case of web plate of larger heights to avoid web buckling.

Elements of Plate Girders

The following are the elements of a typical plate girder:

- Web.

- Flanges.

- Stiffeners.

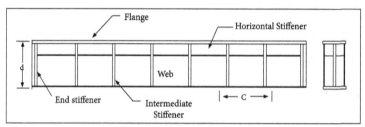

Elements of plate girders.

Webs of required depth and thickness are provided to:

- Keep flange plates at required distances.

- Resist the shear in the beam.

Flanges of required width and thickness are provided to resist bending moment acting on the beam by developing compressive force in one flange and tensile force in another flange.

Stiffeners are provided to safeguard the web against local buckling failure. The stiffeners provided may be classified as:

- Transverse (vertical) stiffeners.

- Longitudinal (horizontal) stiffeners.

Transverse stiffeners are of two types:

- Bearing stiffener.

- Intermediate stiffener.

End bearing stiffeners are provided to transfer the load from beam to the support. At the end certain portion of web of beam acts as a compression member and hence there is possibility of crushing of web. Hence web needs stiffeners to transfer the load to the support. If concentrated loads are acting on the plate girder (may be due to cross beams), intermediate bearing stiffeners are required.

To resist average shear stress, the thickness of web required is quite less. But use of thin webs may result into buckling due to shear. Hence when thin webs are used, intermediate transverse stiffeners are provided to improve buckling strength of web.

Many times longitudinal (horizontal) stiffeners are provided to increase the buckling strength of web. If only one longitudinal stiffeners is provided, it will be at a depth of 0.2 d from the compression flange where 'd' is the depth of web. If another longitudinal stiffener is to be provided it will at mid depth of web.

Web, flange and stiffeners are all plates. They are to be connected suitably by welding to form a single structural system i.e. plate girder. The plate girder has to resist shear force and bending moment acting on it. No plate should fail under any of the designed load.

Design of Components

The complete design of a plate girder consists of the following elements:

- Determination of external loads, estimation of self-weight and computation of S.M. and S.F.
- Determination or fixation of depth of plate girder.
- Design of web plate.
- Design of flanges, including curtailment of flange plates.
- Design of connections between flange angles and web.
- Design of connections between flange plates and flange angles.
- Design of stiffeners.
- Design of web splices.
- Design of flange splices.

3.3.1 Eccentric and Moment Connections

Eccentric Connections

If the load is not passing through the CG of the connection it is called Eccentric Connections. Eg. Bracket connection and seat connection.

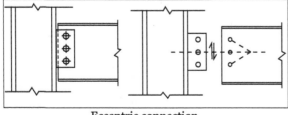

Eccentric connection.

Moment Connections

Moment resisting connections are used in multistorey unbraced buildings and in single storey portal frame buildings. Connections in multistorey frames are most likely to be bolted, full depth end plate connections or extended end plate connections. Where a deeper connection is required to provide a larger lever arm for the bolts, a haunched connection can be used. However, as extra fabrication will result, this situation should be avoided if possible.

For portal frame structures, haunched moment resisting connections at the eaves and apex of a frame are almost always used, as in addition to provide increased connection resistances, the haunch increases the resistance of the rafter. The most commonly used moment resisting connections are bolted end plate beam-to-column connections; these are shown in the figure below.

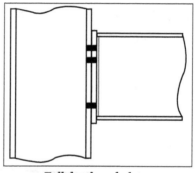

Full depth end plate.

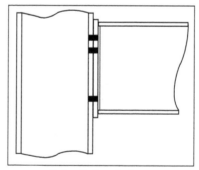

Extended end plate.

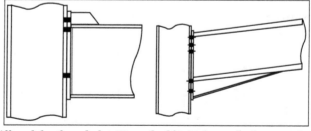

Stiffened depth end plateHaunched beam(may also be extended).

Problems

1. Let us design a welded plate girder of span 24 m carrying super-imposed load of 35 kN/m. Avoid use of bearing and intermediate stiffness. Use Fe 415 (E250) steel.

Solution:

Given:

Span 24 m,

Load of 35 kN/m,

$$\text{Maximum Moment M} = \frac{wL^2}{8} = \frac{58.8 \times 24^2}{8}$$

$$= 4233.6 \text{ kN-m}$$

Maximum shear force = End reaction

$$F = \frac{wL}{2} = \frac{58.8 \times 24}{2} = 705.6 \text{ kN}$$

Depth of Web Plate

If stiffeners are to be avoided,

$$k = \frac{d}{t_w} \le 67$$

$\therefore$ Economical depth of web

$$d = 3\sqrt{\frac{Mk}{fy}} = \left(\frac{4233.6 \times 10^6 \times 67}{250} \right)^{1/3} = 1043 \text{ mm}$$

Use 1000 mm plates,

$$t_w \ge \frac{1000}{67} \text{ i.e., } t_w \ge 14.92$$

Select t_w = 16 mm.

Thus web plate selected is 1000 mm × 16 mm.

Selection of Flange

Neglecting the moment capacity of web, area of flange required is,

$$\frac{A_f f_y d}{1.1} \ge M$$

$$\frac{A_f \times 250 \times 1000}{1.1} \geq 4233.6 \times 10^6$$

To keep the flange in plastic category,

$$\frac{b}{t_f} \leq 8.4$$

Assuming $t_f = 16.8\ t_f$

We get,

$$A_f = 16.8\ t_f\ t_f = 18628$$

$$\therefore t_f = 33.3\ mm$$

Select 40 mm plates. Width of plate require = 18628/40= 465.7

Hence use 480 mm wide and 40 mm thick plates. Section selected is shown in figure.

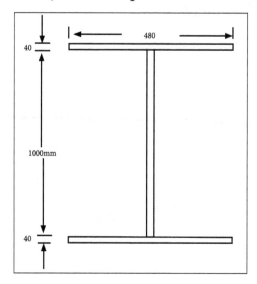

Check for the moment capacity of the girder:

Since it is assumed that only flanges resist the moment and flange is a semi-compact section. Now

$$I_{ZZ} = 2\left[\frac{1}{12} \times 48 \times 40^3 + 480 \times 40 \times \left(\frac{1000 + 2 \times 40}{2}\right)^2\right]$$

$$= 2 \times 5601.28 \times 10^6\ mm^4$$

$$Z_c = \frac{I_{xx}}{y_{max}} = \frac{2 \times 5601.28 \times 10^6}{540} = 20.745 \times 10^6\ mm^3$$

$$M_d = \frac{20.745 \times 10^6 \times 250}{1.1} = 4714.77 \times 10^6 \text{ mm}^3$$

$$= 4714.77 \text{ kN} - \text{m} > M$$

Hence section is adequate.

Shear Resistance of Web

$$V_d = \frac{V_h}{\gamma_{mo}} = \frac{V f_{yw}}{V_{mo}\sqrt{3}} = \frac{dt_w f_{yw}}{V_{mo}\sqrt{3}}$$

$$V_d = \frac{1000 \times 16 \times 250}{1.1\sqrt{3}} = 2099.455 \times 10^3 \text{ N}$$

$$= 2099.455 \text{ kN}$$

Hence section is adequate.

No stiffeners are required.

Check for End Bearing

Bearing strength of web,

$$F_w = (b_1 + n_2) t_2 \frac{f_{yw}}{\gamma_{mo}}$$

Assuming that the width of support is 200 mm, minimum stiff bearing provided by support = 100mm. Dispersion length n_2 = 2.5 × 40 = 100 mm.

F_w = (100 + 100) × 16 × 250/1.1 = 727 × 10³ N = 727 kN > 705.6 kN

Design of Weld Connecting Web Plate and Flange

Maximum shear force = 705.6 kN.

Shear stress in flange at the level of junction of web and flange

$$q = \frac{F}{bl}(\bar{a}_y)$$

$$q = \frac{705.6 \times 10^3}{480 \times 2 \times 5601.28 \times 10^6}\left[480 \times 40 \times \left(500 + \frac{40}{2}\right)\right] = 0.512 \text{ N/mm}^2$$

Shear force per mm length in junction = 0.512 × 480 = 245.76 N

The 40 mm long welds with a gap of 160 mm.

$\therefore$ Final Design = Web: 1000 × 16 mm

Flange : 400 × 40 mm

2. Let us design a reverted plate girder which is simply supported over an effective span of 16 m. It carries a uniformly distributed load of 80 kN/m in addition to its self-weight and two point loads of 400 kN each at 4 m from their supports.

Solution:

Given:

Design Web and Flanges:

Plate Girder - Simply Supported

Effective Span = 16 m

udl = 80 kN/m

2 Point Loads = 400 kN at 4 m from their supports.

Computation of maximum bending Moment and Shear Force:

Factored load on span,

$$=1.5[80 \times 16 + 400 + 400] = 3120 \text{ kN}$$

$$= \frac{3120}{1} \text{N/m} = 3120 \text{ kN/m}$$

Assuming a Self weight of total imposed load,

Self-weight of girder/m $= \dfrac{3120 \times 16}{200} = 249.6\text{k}$

$\therefore$ Maximum B.M $= \dfrac{\omega l^2}{8}$

$$= \frac{249.6 \times 16}{} = 7987 \text{ kN.m}$$

$\therefore$ Maximum S.F $= \dfrac{\omega l}{2} = \dfrac{249.6 \times 16^2}{2} = 1997 \text{ kN}$

Computation of depth of web plate.

Designing the web plate to avoid providing web stiffeners.

$$k = \frac{d}{t_w} \leq 0.67$$

$$d = \left[\frac{MK}{\delta_y}\right]^{1/3} = \left[\frac{7987 \times 10^6 \times 67}{250}\right]^{1/3} = 1288 \text{ mm}$$

Let us provide web plate of depth 1250 mm.

$\therefore$ Thickness of Web, $t_\omega \geq \dfrac{1250}{67} \geq 18.65 \, \text{mm}$

$\therefore$ Let us provide Web Plate of Size 1250 × 20 mm.

Computation of Flange Width

Assuming that the moment capacity of web is neglected, area of Flange required is such that,

$$\frac{A_f f_y d}{\gamma_{m_0}} \geq M$$

$$\frac{A_f \times 250 \times 1000}{1.1} \geq 7987 \times 10^6$$

$$A_f \geq 28114 \, \text{mm}^2$$

Let us provide the flange to be under the semi compact category,

$$\frac{b}{t_f} \leq 13.6 \ t_f = \frac{b}{12}$$

$$A_f = b_f t_f = (12 \, t_f) t_f$$

$$b_f = \sqrt{\frac{28114}{12}}$$

$$t_f = 48.4 \, \text{mm}$$

$$t_f = 50 \, \text{mm}.$$

$$b = 12 \times 50$$

$$= 600 \, \text{mm}$$

Hence Let us use 600 × 50 mm plates for the flanges and 1250 × 20 mm plates for the web.

3. Let us design a riveted plate girder using Fe 415 steel span 22 m to carry a load 25 kN/m.

Solution:

Given:

Permissible bending stress = 165 N/mm²

Shear stress = 100 N/mm²

Total super imposed to girder = 22 × 25 = 550 kN

Self-Wet of girder $= \dfrac{\text{total super imposed} \times \text{span in m load}}{300}$

$$= [(550 \times 22)/300] = 40.33 \text{ kN}$$

Total load = 550 + 41 = 591 kN

$\qquad$ S.F = R_A = R_B = (Total load/2) = (591/2) = 295 say 296 kN

$\qquad$ B.M $= \dfrac{591 \times 22}{8} = 1625.25 \text{ kN}$

Design of the Web

Economic effective depth = $5.5 \ 3\sqrt{\dfrac{M}{f}} = 5.5 \ 3\sqrt{\dfrac{1625.25 \times 10^6}{165}} = 1178.98 \approx 1180 \text{ mm}$

Actual depth = 1180 − 2(5) = 1170 mm

$\qquad$ $h_c = h - 2h_f = 1180 - 2(150) = 880$ mm

Area of Web = s/f_s

$\qquad$ F = 296 kN

$$= \dfrac{296(1000)}{100} = 2960 \text{ mm}^2$$

∴ Thickness of Web $= \dfrac{\text{Area of Web}}{\text{Clear of web}}$

Ratio of Clear Web = 2960/880 = 3.36 = 880/4 = 220

Provide 5 mm

$\qquad$ =880/5 = 176 < 200

∴ Provide Web plate 1180 mm × 5 mm.

Design of Flange,

$\qquad$ $A_f = \dfrac{M}{fh} - \dfrac{1}{8} A_w$

$$= \dfrac{1625.25 \times 10^6}{(165)(1180)} - \dfrac{1}{8}(1180 \times 5) = 7609.95 \approx 7610 \text{ mm}^2$$

Provide 150 mm × 150 mm × 15 flange.

Gross area of two angle = 8566 mm²

20 +1.5 = 21.5mm

Net area provided = 8566 − 4(21.5) (15)

= 7266 mm²

= 7610 − 7266 = 344 mm²

Net width of cover plate = 320 − 2(21.5)

= 277 mm

Thickness of cover plate = 344/277= 1.24 mm

∴ The size of each flange = 320 × 2.5 mm

3.3.1 Roof Trusses

Different Components of Roof Truss

- Top chord members.

- Bottom chord members.

- Struts.

- Slings.

- Sag Tie.

Sag Rod

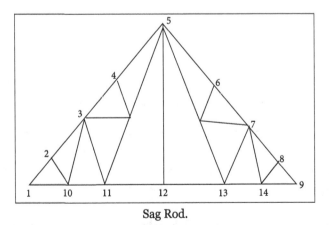

Sag Rod.

In the roof truss shown in figure, member 5-12 is called the sag rod. If no-load is acting at joint 12, the member is not subjected, to any force. Even then this member is provided to reduce the sag in the member 11-13 and hence it is called as sag rod.

General guidelines for fixing spacing of roof stresses:

- 3 to 4.5 m up to l5 m span.
- 4.5 to 6.0 m up to 15 - 30 m span.
- For span more than 40 m, spacing cf 12 - 15 m may be used.

Basis is Live Load Considered in the Design of Roof Truss

Live Load for design of sheets and purling.

Live Load up to $10°$ Slope = 0.75 N/m^2.

For more than $10°$ Slope = $0.75 - 0.02 [\theta - 10]$ where θ is Slope of Sheeting.

However a minimum of 0.4 kN/m^2 as Live Load should be considered in any case.

For the design of trusses the above L.L may be reduced to 2/3 rds.

The Role of End Bearing in Roof Trusses

End bearing are provided at the supports of roof trusses in the form of steel plate anchored on the column in the form of anchor bolts in order to transfer the reaction from the truss member onto the columns.

Roof and Side Coverings

A roof of a building envelope both the covering on the uppermost part of a building or shelter which provides protection from animals and weather notably rain, but also heat, wind and sunlight and the framing or structure which supports the covering.

All the materials laid on the roof frame; includes sheathing, the outer cladding materials, asphalt paper, etc. A roof covering 1 which is not readily flammable and does not slip from position. The following classes have these and additional properties:

Class A is effective against severe fire exposure, does not carry or communicate fire and affords a fairly high degree of fire protection to the roof-deck.

Class B is effective against moderate fire exposure does not readily carry or communicate fire and affords a moderate degree of fire protection to the roof-deck. Class C is effective against light fire exposure does not readily carry or communicate fire and affords a slight degree of fire protection to the roof-deck.

Design of Simple Roof Trusses Involving the Design of Purlins, Members and Joints

Purlins are beams which are provided over roof trusses to support the roof coverings.

Purlins span between the top chords of two adjacent roof trusses. When rafters support the sheeting and rest on purlins then the purlins are placed over the panel points of roof trusses.

Sometimes purlins are placed slightly above the panel points so as to keep the centroids of the purlin sections vertically over the joint center. When sheeting is directly placed over the purlins, these can be placed at intermediate points along the top chord of the trusses.

The top chord of the truss in such a case has a disadvantage as it has to support an axial compressive load along with the reactions from purlins and behaves like a beam column. A channel section is best suited for a purlin. The other types of sections used are an I-section and an angle section. Purlins should be designed carefully as these constitute a large proportion of the steel dead weight in a structure.

Therefore, these affect the overall economy of the structure. Purlins can be designed as simple, continuous or cantilever beams. As per Indian Standard 800: 84, purlins should be designed as continuous beams. But the drawback in designing them like this is that purlins may have to be spliced and continuity cannot be assured through splices. Therefore, the purlin section provided should be as long in length as possible.

The exterior bay width is kept at about 80% of the interior bay width to reduce the maximum moment in the end span of the purlins. This makes the moments in the end span equal to those in the interior spans. Purlins are often designed for only the normal component of forces. The justification given for this is that purlins with roof coverings act like a deep plate girder and due to this the stress from parallel components can be neglected.

Roof Trusses

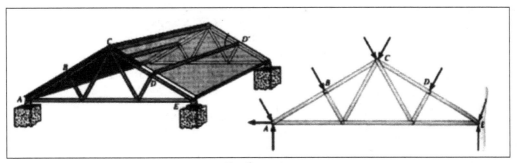

Roof trusses.

A truss is a structure composed of slender members joined together at their end points. If a truss, along with the imposed load, lies in a single plane (as shown at the top right), then it is called a planar truss. A simple truss is a planar truss which begins with a triangular element and can be expanded by adding two members and a joint. For these trusses the number of members (M) and the number of joints (J) are related by the equation $M = 2J - 3$.

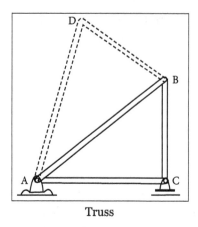

Truss

Selection of Type of Truss

The type of roof truss to be provided depends primarily upon the pitch of the truss. Fink truss, Pratt and Howe truss, Warren truss are provided for large, medium and small pitch, respectively. A skylight can be fitted on them for daylight. When the layout of an industrial building is such that more daylight is required, a North light truss is most suitable, as natural light can be obtained from its geometry.

The pitch of the roof truss is the height of the truss divided by the span. One must not confuse it with the slope of the truss (the slope is numerically twice the pitch). Some of the other factors which may affect the selection of a particular type of roof truss are:

(i) Roof Coverings

The pitch of the truss depends upon the roofing material. The minimum recommended rise of trusses with G.I. sheets is 1 in 6 and with A.C. sheets is 1 in 12.

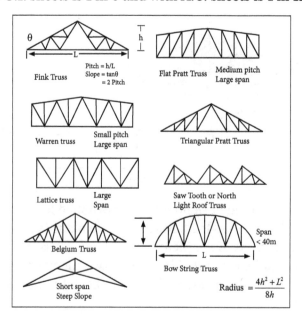

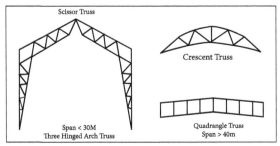

Typical roof truss.

(ii) Fabrication and Transportation

This often guides the type of truss to be selected. Normally trusses are fabricated in the workshop and are transported to the site for erection. From the transportation consideration, depth of the truss becomes a controlling factor as it will not be feasible to transport a very deep truss.

(iii) Aesthetic

From the aesthetic point of view the architect may give a very flat or deep truss, thereby limiting the choice.

(iv) Climate

The climate of a particular area plays an important role in the selection of truss. Drainage of water, ice and snow retention, etc. will have to be given due consideration.

Spacing of Truss

The economic spacing of the truss is the spacing that makes the overall cost of trusses, purlins, roof coverings columns, etc. the minimum. It depends upon the relative cost of the trusses, purlins, roof coverings, spacing of columns, etc. If the spacing of trusses is large, the cost of these trusses per unit area decreases but the cost of purlins increases. If the spacing of trusses is small, the cost of trusses per unit area increases. Roof coverings cost more if the spacing of trusses is large.

Let,

$\qquad$ S = spacing of the trusses.

$\qquad$ t = cost of the trusses/unit area.

$\qquad$ p = cost of the purlins/unit area.

$\qquad$ r = cost of the roof coverings/unit area.

$\qquad$ x = overall cost of the roof system/unit area.

The cost of the truss is inversely proportional to the spacing of trusses,

$$t \propto \frac{1}{S}$$

$$t = \frac{c_1}{S}$$

The cost of the purlins is directly proportional to the square of spacing of trusses,

$$P \propto S^2$$

$$P = c_2 S^2$$

The cost of roof covering is directly proportional to the spacing of trusses,

$$r \propto S$$

$$r = c_3 S$$

$$X = t + p + r$$

$$x \frac{c_1}{S} + c_2 S^2 + c_3 S$$

For the overall cost to be minimum,

$$\frac{dx}{dS} = 0$$

or,

$$\frac{dx}{dS} = 0 = -\frac{c_1}{S^2} + 2c_2 S + c_3$$

or,

$$-\frac{c_1}{S^2} + 2c_2 S + c_3 = 0$$

or,

$$-\frac{c_1}{S} + 2c_2 S^2 + c_3 S = 0$$

or,

$$-t + 2p + r = 0$$

or,

$$t = 2p + r$$

Therefore, for economic spacing of roof trusses, the cost of trusses should be equal to twice the cost of purlins plus the cost of roof coverings. The above expression is used to check the spacing of the roof trusses. It cannot be used to design the spacing as the spacing occurs nowhere in the final equation. As a guide, the spacing of the roof trusses can be kept 1/4 of the span for up to 15 m and 1/5 of span from 15-30 m spans of roof trusses.

Problems

1. A roof term shed is to be build in Lucknow for an industry. The six of shed is 24 m × 40 m. The height of building is 12 m at the ever. Let us determine the basic wind pressure.

Solution:

Given:

Shed = 24 m × 40 m

Height = 12 m

From wind zone map of country V_b = 47 m/sec.

Risk co-efficient k_1 = 1.0

Terrain, height and structure size factor k_2,

k_2 = 0.88 if h = 10 m

= 0.34 if h = 15m

For h = 12 m, k_2 = 0.88 + (0.94 − 0.88) (2/5) = 0.904

Topography factor k_3:

$k_3 = 1 + C_s$,

$c = \dfrac{2}{\varepsilon} = 0$

k_3 = 1.0

Design Wind Speed,

$V_2 = k_1 k_2 k_3 V_b$

V_2 = 1.0 × 0.904 × 1.0 × 47

V_2 = 42.488 m/sec.

Basic Wind Pressure,

$P_z = 0.6 \ V_z^2 = 0.6 \times 42.488^2$

$P_z = 1083 \text{N/m}^2$

$P_z = 1.083 \text{ kN/m}^2$

2. Let us calculate the wind load while designing roof trusses.

Solution:

Wind Load While Design Roof Trusses

Determine Basic Wind Speed [V_b],

Refer IS875 Part 3 - CI.5.2.

Design Wind Speed [V_z],

$$V_z = K_1 K_2 K_3 V_b$$

Where,

K_1 - Risk Co-efficient [Table 1 - IS875 - Part 3]

K_2 - Terrain, height and structure size factors [Table 2-IS875-Part 3]

K_3 -Topography factor. [C1.5.4 of IS875-Part- 3]

Design Wind Pressure [P_z],

$$P_z = 0.6 \, V_z^{\,2}$$

Design Wind Pressure on Roof [F],

Refer IS875 - Part 3 of Cl.6.2.1

$$F = [C_{pe} - C_{\pi}] \, A \, P_d$$

3. Let us determine the driven loads on the purine of an industrial building near Vishakhapatnam given: Class of building: General with life of 50 years, Terrain: Category2, Maximum dimension: 40 m, Width of building: 15m, Height at use level: 8 m, Topography: θ less than 3°.

Solution:

Given:

Permeability: Medium

Span of truss: 15m

Pitch: $\dfrac{1}{5}$

Sheeting: A.C. sheets

Spacing of purlins: 1.35 m

Spacing of trusses: 4 m

1. D.L. Calculations

Weight of sheeting including laps and connectors = 170 N/m²

Self-weight of purlin (assumed) = 100N/m²

Total D.L. on purlin = 170 + 100

$$= 270 \text{ N/m}^2$$

Spacing of purlins = 1.35m

D.L. on purlin = 270 × 1.35 = 364.5 N/m

2. Live Load

Span of truss = 15 m

Pitch = 1/5 Rise = 1/5 × 15 = 3 m

tan θ = 3/7.5 or θ = 621.8°

Live load on purlin = 750 – (21.8 – 10) × 20 = 514 N/m²

Spacing of purlins = 1.35 m

∴ Live load = 514 × 1.35 = 693.9 N/m

3. Wind Load

Basic wind velocity near Vishakhapatnam = 50 m/sec.

$k_1 = 1.0$

k_2 for category 2, class B building with height 8 m, is 0.98.

$k_3 = 1.0$

∴ Design wind pressure p_j = 0.6 × 49² = 1440 N/m²

Wind Pressure Coefficients

$$\frac{h}{w} = \frac{10}{15} = \frac{2}{3}$$

Thus,

$$\frac{1}{2} < \frac{h}{w} < \frac{3}{2}$$

From Table,

When wind angle 0°, for rafter slope 21.8° (wind normal to ridge)

On windward side: $C_{pe} = -0.7 + 1.8/10 \times 0.5 = 0.61$

On lee ward side: $C_{pe} = -0.5$

When wind angle 90°, for rafter slope 21.8° (wind parallel to ridge)

On windward side $C_{pe} = -0.8$

On lee ward side: $C_{pe} = -0.6 - 1.8/10 \times 0.2$

$$= -0.636$$

Internal Wind Pressure Coefficient

For a building with medium permeability,

$$C_{pi} = \pm 0.5$$

$\therefore$ Design wind pressure on windward side,

$$= (-0.8 - 0.5)p_d$$

$$= -1.3 \times 1440 = -1872 \text{ N/m}^2$$

Design wind load on leeward side,

$$= (-0.636 - 0.5) \times 1440 \text{ kN/m}$$

$$= -1635.8 \text{ N/m}^2$$

For Purlin Design

$$DL + LL = 364.5 + 694.9 = 1059.4 \text{ N/m}$$

$$= 1.0594 \text{ kN/m}$$

Wind Load $= -1872 \text{N/m}^2 = -1872 \times 1.35 \text{ N/m}$

$$= -2527.2 \text{ N/m}$$

3. Let us discuss a problem based on the purlin for a roof truss of an industrial building located at Chennai with a span of 15m and length of 60 m. The roofing is AC sheeting. The terrain is an open industrial area. Building is class B building with a clear height of

8m at the caves. And Let us design for 1.5 (DL + WL) combination. Type of truss and roof slope can be assumed.

Solution:

Given:

Span =15m

Length = 60 m

Design of Purlin

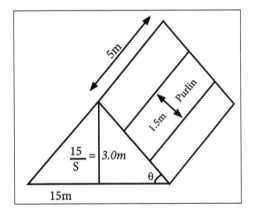

Industrial Building - Chennai.

Span = 15m

Length = 60m

Roofing = AC Sheeting

Terrain = Open

Building = Class B

Height = 8m at Caves

Load Combination = 1.5[DL + WL]

Assume Spacing of truss = 5m c/s

Spacing of Purlin = 1.5 m c/c

Number of Bay = 12 Nos.

Assume Load from roofing material = 200 N/m

Wind Load = 1200 N/m²

Live Load = 600N/m²

Inclination of main rafter = $\tan \theta = 1/5$

$\theta \Rightarrow 11.31°$

Let us assume a live load of 600 N/m²

Step 1:

Minimum depth $= \dfrac{L}{45} = \dfrac{5000}{45} = 111$ mm

Minimum width $= \dfrac{L}{60} = \dfrac{5000}{65} = 83$ mm

Hence Let us try ISA 125 × 95 × 8mm angle section

Step 2: Dead Load and Live Load:

$\quad$ D.L = 200 N/m²

$\quad\quad$ = 200 [Spacing of Purlin]/m

$\quad\quad$ = 200 × 1.5

$\quad$ D.L = 300N/m

$\quad$ L.L = 600 N/m²

$\quad\quad$ = 600[Spacing of Purlin]/m

$\quad\quad$ = 600 [1.5]

$\quad$ L.L = 900 N/m²

$\quad$ Total Load = 1200 N/m

$\quad$ W_u = 1800N/m

Factored Load 1.5 [D.L + L.L] along sheeting,

$\quad$ = 1800cos 11.3°

$\quad$ = 1765 N/m

Factored Load 1.5 [D.L + L.L] normal to sheeting,

$\quad$ = [450 – 1800]

$\quad$ = – 1.350 N/m

∴ DL + LL is critical,

∴ It has a greater value of 1765 N/m

Bending Moment

$$M = \frac{WL^2}{10} = \frac{1765 \times 5^2}{10}$$

$$= 4412.5 \text{ N.m}$$

ISA125 × 95 × 8 mm

$$\frac{b}{t} = \frac{95}{8} = 11.875 \ \& \ \frac{d}{t} = \frac{125}{8} = 15.625$$

Referring Table 2 of IS 800:2007

b/t 10.5 and 15.7

d/t 10.5 and 15.7

It belongs to class 3 semi compact section.

$$M_d = \beta_b Z_P f_y / \gamma_{m_o}$$

$$\beta_b = \frac{Z_e}{Z_P}$$

$$M_d = Z_e f_y / \gamma_{m_o}$$

$$= \frac{30.6 \times 10^3 \times 250}{1.1}$$

$$= 6954545 \text{ N-mm} = 6954.54 \text{N.m} > 44112.5 \text{ N.m}$$

Hence Safe.

4. Let us design an industrial building where the trusses of 16 m span and 4 m rise are spaced at 8 m a part. The building is in medium wind zone in an industrial area of plain land.

Solution:

Given:

Span = 16 m

Spacing of truss = 8 m.

Assume line load = 600 N/m²

c = 1/4 = 14.03°

Step 1:

Trial Section

$$\text{Assuming minimum depth} = \frac{L}{45} = \frac{8000}{45} = 177.78 \text{ mm}$$

$$\text{Minimum width} = \frac{L}{60} = \frac{8000}{66} = 133.33 \text{ mm}$$

Let us try ISA 125 × 95 × 8 mm angle section.

Step 2: Dead load and line load.

Dead load from roofing materials = 200 N/m²

= 200 × 1.5

= 300 N/m

Line load = 600 × 1.5

= 9 ω N/m

Factored (DL + LC) along sheering = 1200 × 1.5 × cos (14°)

= 1746.5 N/m

Normal to sheering = (300 − 1200) (1.5)

= − 1350 N/m

D.L + L.L is critical since it has a greater value 1746 N/n.

Step 3: Bending moment calculations.

$$\text{Maximum bending moment M} = \frac{WL^2}{10}$$

$$= \frac{1746 \times 8^2}{10}$$

= 11174.4 N – m

For the chosen ISA 125 × 95 × 8 mm

$$\frac{b}{t} = \frac{95}{8} = 11.875$$

$$\frac{d}{t} = \frac{125}{8} = 15.625$$

Referring to Table 2 of IS 800:2007.

$$\frac{b}{t} > \ 10.5 \text{ and } 15.7$$

$$\frac{d}{2} > \ 10.5 \text{ and } 15.7$$

$$M_d = \beta_b \, Z_P \, \frac{f_g}{\gamma_{mo}}$$

$$\beta_p = \frac{Z_e}{Z_p}$$

$$M_d = \beta_b \, Z_P \, \frac{f_g}{\gamma_{mo}}$$

$$= Z_e \cdot \frac{fy}{\gamma_{mo}}$$

For ISA 125 × 95 × 8 mm

$$Z_e = 30.6 \times 10^3 \text{ mm}^3$$

$$M_3 = \frac{30.6 \times 10^3 \, (250)}{1.1}$$

$$= 6954545 \text{ N} - \text{mm}$$

$$= 6959.54 \text{ N} - \text{m}$$

$$= 4412.5 \text{ N-m}$$

Hence let us provide ISA 125 × 95 × 8 mm for the piston.

Design of Purlins

Purlins are flexural members used in trusses to support the roof covering and spans between the trusses. Purlins are provided on the top rafter (top chord) at all the joints. The spacing of the purlins depends on the type of the roofing material and for normal materials it ranges from 1.4 to 1.8 m.

The sections used for purlins are usually angles (equal or unequal) as they are economical and variety of sections is available. The new code do not provide the design specifications. Therefore the specifications as per the old code IS: 800 1984 is followed. Cl . 8.9 pp - 69 shall also be followed.

Based on IS : 875 Part 2, LL on inclined roofs shall be taken as,

LL = 0.75 - 0.02 / ° of the slope for slopes > 10° subjected to a minimum of 0.4 kN/m²,

For slopes ≤10°, LL = 0.75 kN/m²

DL of AC sheets = 0.17 kN/m²

And GI sheets = 0.13 kN/m²

Problems

1. Let us discuss about a problem based on a purlin using the following data:

- Spacing of roof trusses = 4.5 m c/c.

- Purlin spacing = 1.8 m c/c.

- Pitch of roof = 1/4.

- Span of the roof = 10 m.

- The vertical load from roof sheets = 180 N/m².

- Wind load intensity normal to roof = 1200 N/m² Use I section.

Solution:

Given:

Design a Purlin:

Spacing of roof truss = 4.5 m c/c

Purlin Spacing = 1.8 m c/c

Pitch of Roof = 1/4

Span of the Roof = 10 m

Vertical load from roof Sheets = 180N/m²

Wind Load intensity normal to roof = 1200 N/m²

Use I section.

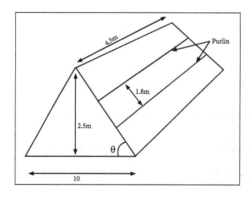

Let us assume a live load of 600 N/m²

Inclination of main rafter $\Rightarrow \tan\theta = \dfrac{1}{4}$

$\qquad \theta = 14.03°$

Resolving Forces Parallel and Perpendicular to Sheeting:

Dead Load of roofing material = 180 N/m²

$\qquad$ Spacing of Purlins = 1.8m

Dead Load = 180 × 1.8 = 324N/m = 0.324kN/m

Assume Live Load = 1.5 kN/m²

$\qquad$ = 1.5 × 1.8 = 2.7 kN/m

Factored Vertical Load = 1.5 [D.L + L.L]

$\qquad$ = 4.536 kN/m

Hence Load acting normal to Sheeting = 4.54 cos 14.03

$\qquad$ = 4.404 kN/m

Load acting parallel to Sheeting = 4.54 sin 14.03

$\qquad$ = 1.1 kN/m

Calculating of B.M and S.F about ZZ-axis:

B.M,

$$M_Z = \frac{W_Z\,l^2}{8} = \frac{14.404 \times 4.5^2}{8} = 11.4 \text{ kN.m}$$

$$M_Y = \frac{W_Y\,l^2}{8} = \frac{101 \times 4.5^2}{8} = 2.78 \text{ kN.m}$$

S.F,

$$F_Z = \frac{11.14 \times 4.5}{2} = 25\,\text{kN}$$

$$F_Y = \frac{2.78 \times 4.5}{2} = 6.25\,\text{kN}$$

Choosing of Trial Section,

Let us try with I SMB 150 section channel.

Section having following properties.

$\qquad r_1 = 9 \text{ mm}, D = 156 \text{ mm}, b = 80 \text{ mm}, t_f = 7.6 \text{ mm}, t_w = 4.8 \text{ mm}$

$Z_{ez} = 96.9 \times 10^3$ mm³

$Z_{ey} = 13.1 \times 10^3$ mm³

$Z_{pz} = 110.5 \times 10^3$ mm³

$d = 150 - (7.6 + 9) = 116.8$ mm

$I_z = 726.4 \times 10^4$ mm⁴

$$Z_{PZ} \text{ required} = \frac{M_z}{f_y}\gamma_{m_0} + 2.5\frac{d}{b}\frac{M_Y}{f_y}\gamma_{m_0}$$

$$= \frac{[11.14]}{250}\times 10^6 \times 10^1 + 2.5\frac{116.8}{80}\times\frac{2.78\times 10^6}{250}\times 1.1$$

$$= 93662.8 \text{ mm}^3 \ < \ 110500 \text{ mm}^3$$

Hence Section is Okay.

Check for Shear,

$$V_{dz} = \frac{f_y/\sqrt{3}}{\gamma_{m_0}} \times \text{Area of web}$$

$$= \frac{250/\sqrt{3}}{101}\times 150 \times 4.8$$

$$= 94478 \text{ N} > 25000\text{N}$$

$$V_{dy} = \frac{f_y/\sqrt{3}}{\gamma_{m_0}} \times \text{Area of Flange}$$

$$= \frac{250/\sqrt{3}}{1.1}\times 2 \times b \times t_f$$

$$= \frac{250}{\sqrt{3}\times 1.1}\times 2 \times 80 \times 7.6$$

$$= 159563 \text{ N} > 6250 \text{ N}$$

Hence the section is adequate in shear.

Calculation of Design Capacity of the Section

$$b/t_f = \frac{80}{7.6} = 10.526 > 9.4\varepsilon < 10.5\varepsilon$$

$$b/t_w = \frac{116.8}{4.6} = 24.33 < 84\varepsilon$$

It belongs to compact section, $\beta_b = 1.0$.

$$M_{d_z} = \beta_b Z_p f_y / \gamma m_o < 1.2 \ Z_e f_y / \gamma_{mo}$$

$$= \frac{1 \times 110.5 \times 10^3 \times 2.5}{1.1} < \frac{1.2 \times 96.9 \times 10^3 \times 250}{1.1}$$

$$M_{d_z} = 25.11 \times 10^6 \ N - mm < 26.43 \times 10^6 \ N - mm$$

$$M_{d_z} = \frac{f_y \ Z_{PY}}{\gamma m_o} \left[\because Z_{PY} = bt_f \frac{b}{2} \right]$$

$$= 80 \times 7.6 \times \frac{80}{2}$$

$$= 24320 \ mm^3$$

$$= \frac{250 \times 24320}{1.1} = 5.527 \times 10^6 \ N - mm$$

Check Whether

$$\frac{M_z}{M_{d_z}} + \frac{M_y}{M_{d_y}} \leq 1.0$$

$$\frac{11.14 \times 10^6}{25.11 \times 10^6} + \frac{2.78 \times 10^6}{5.53 \times 10^6} \leq 1.0$$

$$0.94 < 1.0$$

Hence Safe.

Check for deflection of purlin:

$$I_z = 726.4 \times 10^4 \ mm^4. \ L = 4500 \ mm$$

$$\omega = 4.54 \ kN/m = 4.54 \ N/mm. \ E = 2 \times 10^5 \ MPa.$$

$$\delta = \frac{5}{384} \times \frac{10L^4}{EI} < \frac{L}{150} \ [\text{Permissible deflection}]$$

$$= \frac{5}{384 \times 4.54 \times 4500^4 < 4500 - 1 \ \lambda} = 16.69 < 30 \ mm$$

2. Let us design a channel section purlin for the following data; Spacing of trusses = 4.2 m Spacing of purlins = 2 m.

Solution:

Given:

Spacing of trusses = 4.2 m Spacing of purlins = 2 m.

Line load on galvanized iron roofing sheet = 0.6 kN/m².

Wind load =1.4 kN/m².

Slope of main = 31°.

Step 1: Resolving forces parallel and perpendicular to sheeting.

Assume dead load = 200 N/m²

Let the self-weight of purlins = 125 N/m³

Total dead load = 200 + 125

= 325 N/m²

= 0.325 kN/m²

Live load = 0.6 kN/m².

Factored DL + LL is given as:

= 1.5 (0.325 + 0.6)

= 1.3875 kN/m²

Spacing of purlin = 2 m.

Load acting vertically downward:

W = (1.3875) (2)

W = 2.775 kN/m

Hence load acting normal to sheeting,

W_x = 2.775 cos θ

= 2.775 cos 30°

W_x = 2.403 kN/m

Load acting parallel to sheeting,

W_y = 2.775 sin θ

W_y = 1.3875 kN/m

Step 2: Calculation of BM and SF about Z – Z axis.

Bending moments,

$M_z = l^2/8$

$$M_y = l^2/8$$

1-span of purlin = 4.2 m.

$$M_z = \frac{(2.403)(4.2)^2}{8} = 5.298\,kN - m$$

$$M_y = \frac{(1.3875)(4.2)^2}{8} = 3.059\,kN - m$$

Shear Forces,

$$F_z = \frac{(2.403)(4.2)}{2} = 5.0463\,kN$$

$$F_y = \frac{(1.3575)(4.2)}{2} = 2.91375\,kN$$

Step 3: Choosing of total section.

Let us use ISMC 125 channel section.

$$Z_{ez} = 66.6 \times 10^3 \text{ mm}^3$$

$$Z_{ey} = 13.1 \times 10^3 \text{ mm}^3$$

$$t_f = 8.1 \text{ mm}$$

$$t_w = 5.0 \text{ mm}$$

$$d = 125 - 2(8.1)$$

$$= 108.8 \text{ m}$$

$$b = 65 \text{ mm}$$

$$Z_{P2} = \frac{m_z}{fy}\gamma_{mo} + 2.5\bar{b}\frac{M_y}{fy}\gamma_{mo}$$

$$= \frac{5.298 \times 10^6}{250}(1.1) + 2.5\left(\frac{108.8}{65}\right)\left(\frac{3.059 \times 10^6}{250}\right)(1.1)$$

$$= 23311.2 + 56323.2$$

$$= 79634.4 \text{ mm}^3$$

$$Z_{p2} \text{ of ISMC 125} = 79.63 \times 10^3 \text{ mm}^3$$

Hence adequate.

Step 4: Check for Shear.

$$V_{dz} = \frac{f_y/\sqrt{3}}{\gamma_{mo}} \times \text{Area of web}$$

$$= \left(\frac{250/\sqrt{3}}{1.1}\right)(108.8 \times 5) - 71384 = 71.38 \text{ kV ZF}_2.$$

$$V_{dy} = \frac{f_y/\sqrt{3}}{\gamma_{mo}} \times \text{Area of Flanges}$$

$$= \frac{250\sqrt{3}}{1.1}(2 \times 65 \times 8.1)$$

$$= 138174 = 138.17 \text{ kN F}_y$$

Hence the section is adequate.

Step 5: let us Calculation of design capacity of the section.

Classification of section $\dfrac{b}{t_s} = \dfrac{65}{8.1} = 8.025$

$$\frac{d}{t_w} = \frac{108.8}{5} = 2.176$$

IS 800:2007 $\dfrac{b}{t_5} < 9.4$

The section classification is plastic

$$\frac{b}{t_w} < 4.2$$

$$M_{dz} = \beta_b Z_p f_y / \gamma_{mo}$$

For plastic section $\beta_b = 1.0$

$$M_{dz} = \frac{\beta_b f_y Z_{py}}{\gamma_{mo}} = \frac{(1.0)(250)(77.2 \times 10^3)}{1.1}$$

$$= 17545454 \text{ N-mm}$$

$$\approx 17.54 \times 10^6 \text{ N-mm}$$

$$\frac{1.2 Z_{ez} f_y}{\gamma_{mo}} = \frac{1.2(66.6 \times 10^3)(250)}{1.1}$$

$$= 18163636 \text{ N-mm}$$

$$= 17.54 \times 10^6 \text{ N-mm } (M_z)$$

$$M_{dz} = 17.54 \times 10^6 \text{ N-mm}$$

$$M_{dy} = \frac{f_y Z_{py}}{\gamma_{mo}}$$

Where,

$$Z_{py} = b\, t_f\, (b/2)$$

$$= \frac{b^2}{2} t_f$$

$$= \frac{(65)^2 (8.1)}{2}$$

$$= 17.111 \text{ mm}^3$$

$$M_{dy} = \frac{250(17111)}{1.1} = 3888920 \text{ N-mm}$$

$$\frac{1.5 Z_{zy} f_y}{\gamma_{mo}} = \frac{1.5(13.1)\times10^3 (250)}{1.1}$$

$$= 4465909 \text{ N-mm}$$

$$= 3888920 \text{ N-mm}$$

$$M_{dy} = 3888920$$

$$= 3.889 \times 10^6 \text{ N-mm}$$

Step 6: Check Weather

$$\frac{M_2}{M_{dz}} + \frac{M_y}{M_{dy}} \leq 1.0$$

$$M_z = 4.806 \times 10^6 \text{ N-mm}$$

$$M_y = 2.775 \times 10^6 \text{ N-mm}$$

$$\frac{4.806\times10^6}{17.545\times10^6} + \frac{2.775\times10^6}{3.889\times10^6} = 0.9876 < 1$$

Hence the section is adequate.

Step 7: Check for deflection

$$1_z \text{ for ISMC } 125 = 416.4 \times 10^4 \text{ mm}^4$$

$$W = 2.775 \text{ kN/n} = 2.775 \text{ N/mm}$$

Actual deflection is given by,

$$\delta = \frac{5\,Wl^4}{384\,EI} = \frac{5}{384}\left(\frac{2.775 \times (4000)^4}{(2\times10^5)(416.4\times10^4)}\right)$$

$$E = 2 \times 10^5 \text{ MPa}$$

$$= 11.107 \text{ mm}$$

As per table 6 of IS 800 : 2007.

Permissible deflection $=L/150$

$$= \frac{4000}{150} = 2.6467 \text{ mm}$$

$$= 11.107 \text{ mm}$$

Step 8:

Factored DL = (Load Factor) (Total dead load) (Spacing of listen,

$$= (1.5)\,(325)\,(2.1)$$

$$= 1023.75 \text{ N/m}$$

$$= 1.02 \text{ kN/m}$$

Wind load = 1.5 kN/m²

$$= (1.5)\,(2)$$

$$= 3 \text{ kN/m}$$

Factored wind force = 1.5 (3)

$$= 4.5 \text{ kN/m}$$

Load normal to sheeting = $- 4.5$ + factored DL

$$= - 4.5 + 1.02 \cos 30°$$

$$= - 3.66 \text{ kN/m (outward)}.$$

Load parallel to sheeting = $0.975 \sin 30°$

$$= 0.5200 \text{ kN/m}$$

$$M_{zz} = \frac{3.66(4)^2}{8} = 7.32 \text{ kN/m}$$

$$M_{yy} = \frac{0.525(4)}{8} = 1.09 \text{ kN/m}$$

$$\frac{M_z}{M_{dz}} + \frac{M_y}{M_{dy}} \text{ should be} \le 1.0$$

$$\frac{7.32 \times 10^6}{7.382 \times 10^6} \text{ should be} \le 1.0$$

0.9916 + 0.25 = 1.24 > 1

Hence the section may be reversed by selecting next suitable channel section with greater γ_{yy}.

3. Let us design a purlin based on the purlin for the following specifications:

Span of Truss = 12mc/c

Pitch = 1/5 of span

Spacing of Truss = 5 m c/c

Spacing of purlin = 1.5mc/c

Load from roofing materials etc. = 200 N/m²

Wind load = 1200 N/m²

Use angle section.

Solution:

Given:

Consider 1 meter run of the purlin loads,

Dead load of forcing material = 1.5 × 100 = 240 N/m

Dead load of purlin (assumed) = 120 N/m

Total DL = 360 N/m

Wind load acting normal to rag,

= 1200 × 1.5

= 1800 N/m

Using angle section,

Total dead load = 360 N/m

Wind load = 1800 N/m

Total load = 1800 + 360 = 2160 N/m

Maximum bending moment $= M = \dfrac{2160 \times 3.5^2}{10}$

M = 2646 N/m

Section modulus required $= Z = \dfrac{M}{f} = \dfrac{26469 \times 10^3}{165}$

Z = 16036 mm³

Minimum depth of purlin $= \dfrac{\text{Span of purlin}}{60}$

$= \dfrac{3.5 \times 10^3}{60} = 58.3 \, \text{mm}$

Minimum width of purlin,

$= \dfrac{\text{Span of purlin}}{45}$

$= \dfrac{3.50 \times 10^3}{45}$

= 77.8 mm

Provide 90 mm × 60 mm × 10 mm angle having a section modulus of 18600 mm³.

Permissions

Index

Printed in the USA
CPSIA information can be obtained
at www.ICGtesting.com
JSHW051751040324
58547JS00005B/110